■ ■ ■ 智能系统与技术丛书

Natural Language Processing with Python Cookbook

自然语言处理Python进阶

克里希纳·巴夫萨（Krishna Bhavsar）
[印度] 纳雷什·库马尔（Naresh Kumar） 著
普拉塔普·丹蒂（Pratap Dangeti）

陈钰枫 译

机械工业出版社
China Machine Press

图书在版编目（CIP）数据

自然语言处理 Python 进阶 /（印）克里希纳 · 巴夫萨（Krishna Bhavsar）等著；陈钰枫译．—北京：机械工业出版社，2019.1
（智能系统与技术丛书）
书名原文：Natural Language Processing with Python Cookbook

ISBN 978-7-111-61643-6

I. 自…　II. ①克…　②陈…　III. 软件工具 - 程序设计　IV. TP311.561

中国版本图书馆 CIP 数据核字（2019）第 003846 号

本书版权登记号：图字　01-2018-1366

Krishna Bhavsar, Naresh Kumar, Pratap Dangeti: Natural Language Processing with Python Cookbook (ISBN: 978-1-78728-932-1).

自然语言处理 Python 进阶

出版发行：机械工业出版社（北京市西城区百万庄大街 22 号　邮政编码：100037）
责任编辑：赵　静　　责任校对：李秋荣
印　　刷：三河市宏图印务有限公司　　版　　次：2019 年 2 月第 1 版第 1 次印刷
开　　本：186mm × 240mm　1/16　　印　　张：13.75
书　　号：ISBN 978-7-111-61643-6　　定　　价：59.00 元

凡购本书，如有缺页、倒页、脱页，由本社发行部调换
客服热线：（010）88379426　88361066　　投稿热线：（010）88379604
购书热线：（010）68326294　88379649　68995259　　读者信箱：hzit@hzbook.com

THE TRANSLATOR'S WORDS

译者序

当第一次阅读本书的时候，我深感它就是目前寻求自然语言处理和深度学习入门及进阶方法的读者所需要的。

眼下自然语言处理在人工智能世界大放异彩，渴望徜徉其中的读者众多。我在自然语言处理方面有研究基础，此前也翻译过一些该领域的前沿文章，希望这份译本能帮助想通过 Python 工具深入钻研自然语言处理的读者。

本书是利用 Python 语言解决自然语言处理多种任务的应用指南，旨在帮助读者快速浏览自然语言处理的全貌，进而掌握自然语言处理各项任务的基本原理，最终快速踏入实践阶段。Python 编程语言以其清晰简洁的语法、易用性和可扩展性以及丰富庞大的库深受广大开发者的喜爱。Python 内置了非常强大的机器学习代码库和数学库，使它理所当然地成为自然语言处理的开发利器。而其中的 NLTK 是使用 Python 处理自然语言数据的领先平台。它为各种语言资源（比如 WordNet）提供了简便易用的界面，它还能用于高效完成文本预处理、词性标注、信息抽取和文本分类等多种自然语言处理任务。

值得一提的是，本书特别涵盖了自然语言处理的高阶任务，比如主题识别、指代消解和创建聊天机器人等。本书最后两章介绍了深度学习在自然语言处理中的应用。每一节都大致包含三个部分：准备工作、如何实现和工作原理，对自然语言处理各种任务的方方面面都有所涉及。这意味着，如果读者能够踏实地完成本书中给出的所有实例，就能切实掌握自然语言处理各种任务的基础技术和原理方法，为进一步学习和研究奠定基础。换言之，深入阅读本书，对读者探索深度学习技术在自然语言处理中的应用会有极大帮助。

最后，感谢机械工业出版社华章公司的编辑，是他们的鼓励和支持使得本书中文版能够与读者见面。感谢武文雅和李兴亚等多位研究生的辅助和校对，感谢我的家人的支持。尽管我们努力准确地表达出作者介绍的思想和方法，但仍难免有不当之处。若发现译文中的错误，敬请指出，我们将非常感激，请将相关意见发往 chenyf@bjtu.edu.cn。

陈钰枫

2018 年 11 月

PREFACE

前　言

亲爱的读者，感谢你选择本书来开启你的自然语言处理（Natural Language Processing, NLP）之路。本书将从实用的角度带领你由浅入深逐步理解并实现 NLP 解决方案。我们将从访问内置数据源和创建自己的数据源开始指引你踏上这段旅程。之后你将可以编写复杂的 NLP 解决方案，包括文本规范化、预处理、词性标注、句法分析等。

在本书中，我们将介绍在自然语言处理中应用深度学习所必需的各种基本原理，它们是目前最先进的技术。我们将使用 Keras 软件来讨论深度学习的应用。

本书的出发点如下：

- 内容设计上旨在通过细节分析来帮助新手迅速掌握基本原理。并且，对有经验的专业人员来说，它将更新各种概念，以便更清晰地应用算法来选择数据。
- 介绍了在 NLP 中深度学习应用的新趋势。

本书的组织结构

第 1 章教你使用内置的 NLTK 语料库和频率分布。我们还将学习什么是 WordNet，并探索其特点和用法。

第 2 章演示如何从各种格式的数据源中提取文本。我们还将学习如何从网络源提取原始文本。最后，我们将从这些异构数据源中对原始文本进行规范并构建语料库。

第 3 章介绍一些关键的预处理步骤，如分词、词干提取、词形还原和编辑距离。

第 4 章介绍正则表达式，它是最基本、最简单、最重要和最强大的工具之一。在本章中，你将学习模式匹配的概念，它是文本分析的一种方式，基于此概念，没有比正则表达式更方便的工具了。

第 5 章将学习如何使用和编写自己的词性标注器和文法规则。词性标注是进一步句法分析的基础，而通过使用词性标记和组块标记可以产生或改进文法规则。

第 6 章帮助你了解如何使用内置分块器以及训练或编写自己的分块器，即依存句法分析器。在本章中，你将学习评估自己训练的模型。

第 7 章介绍信息抽取和文本分类，告诉你关于命名实体识别的更多信息。我们将使用内置的命名实体识别工具，并使用字典创建自己的命名实体。我们将学会使用内置的文本分类算法和一些简单的应用实例。

第 8 章介绍高阶自然语言处理方法，该方法将目前为止你所学的所有课程结合到一起，并创建应对你现实生活中各种问题的适用方法。我们将介绍诸如文本相似度、摘要、情感分析、回指消解等任务。

第 9 章介绍深度学习应用于自然语言处理所必需的各种基本原理，例如利用卷积神经网络（CNN）和长短型记忆网络（LSTM）进行邮件分类、情感分类等，最后在低维空间中可视化高维词汇。

第 10 章描述如何利用深度学习解决最前沿的问题，包括文本自动生成、情景数据问答，预测下一个最优词的语言模型以及生成式聊天机器人的开发。

本书需要你做什么

为了成功完成本书的实例，你需要在 Windows 或 Unix 操作系统上安装 Python 3.x 及以上版本，硬件要求：CPU 2.0GHz 以上，内存 4GB 以上。就 Python 开发的 IDE 而言，市场上有许多可用的 IDE，但我最喜欢的是 PyCharm 社区版。它是一款由 JetBrains 开发的免费开源工具，它的技术支持很强大，会定期发布该工具的升级和修正版本，你只要熟悉 IntelliJ 就能保持学习进度顺畅。

本书假设你已经了解 Keras 的基本知识和如何安装库。我们并不要求读者已经具备深度学习的知识和数学知识，比如线性代数等。

在本书中，我们使用了以下版本的软件，它们在最新的版本下都能很好地运行：

- Anaconda 3　4.3.1（Anaconda 中包括所有 Python 及相关包，Python　3.6.1, NumPy　1.12.1, pandas　0.19.2）
- Theano　0.9.0
- Keras　2.0.2
- feedparser　5.2.1
- bs4　4.6.0
- gensim　3.0.1

本书的读者对象

本书适用于想利用 NLP 提升现有技能来实现高阶文本分析的数据科学家、数据分析师和数据科学专业人员，建议读者具备自然语言处理的一些基本知识。

本书也适用于对自然语言处理知识毫无了解的新手，或是希望将自己的知识从传统的 NLP 技术扩展到最先进的深度学习应用技术的有经验的专业人士。

小节

在本书中，有几个标题经常出现（准备工作、如何实现、工作原理、更多、参见）。为了明确说明如何完成一个实例，本书使用了如下内容排布：

准备工作

本节介绍完成实例的预期结果，并说明如何安装所需的软件或初步设置。

如何实现

本节包含完成实例所需遵循的步骤。

工作原理

本节通常对上一节的操作进行详细的解释。

更多

本节包含实例的补充信息，以便读者对实例的实现方法有更多的了解。

参见

本节为实例的其他有用信息提供有效的链接。

约定

在本书中，你会发现许多不同类型信息的文本格式。下面是这些格式的一些范例和它们的含义解释。

任何命令行的输入或输出格式如下：

```
# Deep Learning modules
>>> import numpy as np
>>> from keras.models import Sequential
```

新术语和重要词以加粗字体显示。

警告或重要提示将跟在这样的符号后面。

提示或小技巧将跟在这样的符号后面。

读者反馈

欢迎读者的反馈。让我们知道你对本书的看法，哪些部分喜欢还是不喜欢。读者反馈对我们来说很重要，因为它有助于我们了解读者真正能从中获益最多的地方。将你的反馈发送到邮箱 feedback@packtpub.com 即可，并且在你的邮箱标题中提到本书的书名。如果你有一个擅长或是比较感兴趣的话题，可以撰写或投稿书刊，请在 www.packtpub.com/authors 网站上查看作者指南。

客户支持

既然你是一本 Packt 书的拥有者，在购买时能够获得很多额外的资源。

示例代码下载

你可以从 http://www.packtpub.com 通过个人账号下载本书的示例代码文件。如果你在别处购买了本书，可以访问 http:// www. packtpub.com/support 并进行注册，就可以直接通过邮件方式获得相关文件。你可以按照以下步骤下载代码文件：

1. 使用你的电子邮件地址和密码登录或注册到我们的网站；
2. 将鼠标指针置于顶部的 SUPPORT 选项卡上；
3. 点击 Code Downloads & Errata；
4. 在 Search 框中输入图书的名称；
5. 选择你要下载的代码文件的相应图书；
6. 从你购买本书的下拉菜单中选择；
7. 点击 Code Download。

你也可以在 Packt 出版社网站关于本书的网页上通过点击 Code Files 按钮来下载代码文件。通过在 Search 框中输入图书的名称来访问该页面。请注意，你需要登录到你的 Packt 账户。一旦文件被下载，请确保你使用的是最新版本的工具解压文件夹：

- WinRAR/7-Zip for Windows
- Zipeg/iZip/UnRarX for Mac
- 7-Zip/PeaZip for Linux

本书的代码可从 GitHub 下载，托管在 https://github.com/PacktPublishing/Natural-Language-Processing-with-Python-Cookbook 上。其余代码包可以从 https://github.com/PacktPublishing/ 上丰富的图书和视频目录中获取。请检验测试！

ABOUT THE AUTHORS

作者简介

克里希纳·巴夫萨（Krishna Bhavsar）花了大约10年时间在各行业领域如酒店业、银行业、医疗行业等进行自然语言处理、社交媒体分析和文本挖掘方面的研究。他致力于用不同的NLP语料库如Stanford CoreNLP、IBM的SystemText和BigInsights、GATE和NLTK来解决与文本分析有关的行业问题。克里希纳还致力于分析社交媒体给热门电视节目和流行零售品牌以及产品带来的效应。2010年，他在NAACL上发表了一篇关于情感分析增强技术的论文。近期，他创建了一个NLP管道/工具集并开源以便公众使用。除了学术和科技，克里希纳还热衷于摩托车和足球，空闲时间喜欢旅行和探索。他骑摩托车参加过环印度公路旅行并在东南亚和欧洲大部分国家徒步旅行过。

首先，我要感谢我的母亲，她是我生命中最大的动力和坚强的支柱。我要感谢Synerzip的管理团队和我所有的朋友在我写作过程中对我的支持。最后，特别感谢Ram和Dorothy在这困难的一年让我保持前进。

纳雷什·库马尔（Naresh Kumar）曾为财富500强企业设计、实施和运行超大型因特网应用程序，在这方面他拥有超过十年的专业经验。他是一位全栈架构师，在电子商务、网络托管、医疗、大数据及分析、数据流、广告和数据库等领域拥有丰富的实践经验。他依赖开源并积极为其做贡献。纳雷什一直走在新兴技术的前沿，从Linux系统内部技术到前端技术。他曾在拉贾斯坦邦的BITS-Pilani学习，获得了计算机科学和经济学的双学位。

普拉塔普·丹蒂（Pratap Dangeti）在班加罗尔的研究和创新实验室开发机器学习和深度学习方法，以用于结构化、图像和TCS文本数据。他在分析和数据科学领域拥有丰富的经验，并在IIT Bombay获得了工业工程和运筹学项目的硕士学位。普拉塔普是一名人工智能爱好者。闲暇时，他喜欢阅读下一代技术和创新方法。他还是Packt出版的《Statistics for Machine Learning》一书的作者。

我要感谢我的母亲Lakshmi，感谢她在我的职业生涯以及撰写本书的过程中给予我的支持。我要把这本书献给她。我还要感谢我的家人和朋友，没有他们的鼓励，我不可能完成这本书。

ABOUT THE REVIEWER

审校者简介

Juan Tomas Oliva Ramos 是墨西哥瓜纳华托大学的环境工程师，他拥有行政工程和质量硕士学位。他在专利管理和开发、技术创新项目以及通过流程统计控制开发技术解决方案方面拥有超过 5 年的经验。

自 2011 年以来，Juan Tomas Oliva Ramos 一直是统计、创业和项目技术开发方面的教师。他还是一名企业家导师，并在墨西哥瓜纳华托德尔林孔理工高等研究院开设了一个新的技术管理和创业部门。

Juan 是 Alfaomega 的审稿人，致力于完成《Wearable Designs for Smart Watches, Smart TVs and Android Mobile Devices》一书。

Juan 还通过编程和自动化技术开发了原型系统，用于改进操作，并已经注册了专利。

感谢 Packt 出版社让我有机会审校这本精彩的书，并与一群志同道合的伙伴合作。

感谢我美丽的妻子 Brenda，我们的两位魔法公主（Maria Regina 和 Maria Renata）和我们的下一位成员（Angel Tadeo），你们每天都给了我力量、幸福和欢乐。谢谢你们。

CONTENTS

目　录

CHAPTER 1

第 1 章

语料库和 WordNet

1.1 引言

解决任何实际的**自然语言处理（NLP）**问题，都需要处理大量的数据。这些数据通常以公开语料库的形式存在，并可以由 NLTK 数据包的附加组件提供。例如，如果要创建一个拼写检查器，需要用一个大型单词语料库进行匹配。

本章将涵盖以下内容：

- 介绍 NLTK 提供的各种有用的文本语料库
- 如何用 Python 访问内置语料库
- 计算频率分布
- WordNet 及其词法特征介绍

我们将通过实践的方式来理解这些内容。下面我们会进行一些练习，通过实例来完成这些学习目标。

1.2 访问内置语料库

如前所述，NLTK 有许多可供使用的语料库。这里假设你已经在计算机上完成了 NLTK 数据库的下载和安装。如果没有，你可以通过网址 http://www.nltk.org/data.html 下载。此外，NLTK 数据库内的完整语料库列表可以通过网址 http://www.nltk.org/nltk_data/ 获取。

现在，我们的第一个任务 / 实例是学习如何访问这些语料库。我们在路透社语料库（Reuters corpus）上做一些实验。将语料库导入我们的程序中，并尝试用不同的方式进行访问。

如何实现

1. 创建一个新文件，将其命名为 reuters.py，并在该文件中添加以下代码。这是在整个

NLTK 数据集中仅访问路透社语料库的特定方式：

```
from nltk.corpus import reuters
```

2. 当我们想知道这个语料库中有什么内容时，最简单的方法是调用语料库对象中的 fileids() 函数。在程序中添加以下代码：

```
files = reuters.fileids()
print(files)
```

3. 运行该程序，将得到如下输出：

```
['test/14826', 'test/14828', 'test/14829', 'test/14832',
'test/14833', 'test/14839',
```

这些是路透社语料库中的文件列表和它们的相对路径。

4. 访问这些文件的具体内容。使用语料库对象的 words() 函数来访问 test/16097 文件：

```
words16097 = reuters.words(['test/16097'])
print(words16097)
```

5. 再次运行该程序，会出现一行新的输出内容：

```
['UGANDA', 'PULLS', 'OUT', 'OF', 'COFFEE', 'MARKET', ...]
```

输出了 test/16097 文件中的单词列表。虽然整个单词列表被加载到内存对象中，此处仅输出部分结果。

6. 从 test/16097 文件中获取特定数量的单词（例如 20 个）。当然，我们可以指定想要获取的单词数，并将其存储在列表中以供使用。添加如下两行代码：

```
words20 = reuters.words(['test/16097'])[:20]
print(words20)
```

运行代码，输出结果如下：

```
['UGANDA', 'PULLS', 'OUT', 'OF', 'COFFEE', 'MARKET', '-', 'TRADE',
'SOURCES', 'Uganda', "'", 's', 'Coffee', 'Marketing', 'Board', '(',
'CMB', ')', 'has', 'stopped']
```

7. 进一步，路透社语料库不仅仅是一个文件列表，而且还被按层次分成 90 个主题。每个主题都有许多与之关联的文件。也就是说，当你访问任何一个主题时，实际上访问的是与该主题相关的所有文件的集合。添加如下代码以输出主题列表：

```
reutersGenres = reuters.categories()
print(reutersGenres)
```

运行代码，输出控制台将有如下输出：

```
['acq', 'alum', 'barley', 'bop', 'carcass', 'castor-oil', 'cocoa',
'coconut', 'coconut-oil', ...
```

显示了所有的 90 个类别。

8. 最后，编写四行简单的代码，这不仅可以访问两个主题，还可以将单词以一行一个句子这样松散的方式打印出来。将以下代码添加到 Python 文件中：

```
for w in reuters.words(categories=['bop','cocoa']):
  print(w+' ',end='')
  if(w is '.'):
    print()
```

9. 简单解释一下，我们首先选择了类别 bop 和 cocoa，并打印这两个类别文件中的每个单词。每遇到一个点号（.），就插入一个新的行。运行代码，控制台将输出以下内容：

```
['test/14826', 'test/14828', 'test/14829', 'test/14832',
'test/14833', 'test/14839', ...
['UGANDA', 'PULLS', 'OUT', 'OF', 'COFFEE', 'MARKET', ...]
['UGANDA', 'PULLS', 'OUT', 'OF', 'COFFEE', 'MARKET', '-', 'TRADE',
'SOURCES', 'Uganda', "'", 's', 'Coffee', 'Marketing', 'Board', '(',
'CMB', ')', 'has', 'stopped']
['acq', 'alum', 'barley', 'bop', 'carcass', 'castor-oil', 'cocoa',
'coconut', 'coconut-oil', ...
SOUTH KOREA MOVES TO SLOW GROWTH OF TRADE SURPLUS South Korea ' s
trade surplus is growing too fast and the government has started
taking steps to slow it down , Deputy Prime Minister Kim Mahn-je
said .
He said at a press conference that the government planned to
increase investment , speed up the opening of the local market to
foreign imports, and gradually adjust its currency to hold the
surplus " at a proper level ." But he said the government would not
allow the won to appreciate too much in a short period of time .
South Korea has been under pressure from Washington to revalue the
won .
The U .
S .
Wants South Korea to cut its trade surplus with the U .
S ., Which rose to 7 .
4 billion dlrs in 1986 from 4 .
3 billion dlrs in 1985 .
.
.
.
```

1.3 下载外部语料库，加载并访问

现在我们已经学会了如何加载和访问内置语料库，下面将学习如何下载并加载，以及访问外部语料库。许多内置语料库都非常适用于训练，但是为了解决实际问题，通常需要一个外部数据集。在本节实例中，我们将使用 Cornell CS 电影评论语料库，该语料库对评论做了正面和负面的标记，已被广泛应用于训练情感分析模块。

1.3.1 准备工作

首先，你需要从互联网上下载数据集并将其解压缩。链接如下：http://www.cs.cornell.edu/people/pabo/movie-review-data/mix20_rand700_tokens_cleaned.zip。然后，将生成的评

论（Reviews）目录存储在计算机的安全位置。

1.3.2 如何实现

1. 创建一个名为 external_corpus.py 的新文件，并向该文件添加如下内容：

```
from nltk.corpus import CategorizedPlaintextCorpusReader
```

由于下载的语料库已经分类，我们将使用 CategorizedPlaintextCorpusReader 来读取和加载所给的语料库。用这种方式来获取正面评论和负面评论。

2. 读取语料库。我们需要知道从 Cornell 上下载的文件解压缩后的 Reviews 文件夹的绝对路径，添加以下四行代码：

```
reader = CategorizedPlaintextCorpusReader(r'/Volumes/Data/NLP-
CookBook/Reviews/txt_sentoken', r'.*\.txt', cat_pattern=r'(\w+)/*')
print(reader.categories())
print(reader.fileids())
```

第一行是通过调用 CategorizedPlaintextCorpusReader 构造函数来读取语料库。从左到右的三个参数分别是计算机上 txt_sentoken 文件夹的绝对路径，txt_sentoken 文件夹中的所有示例文档名以及给定语料库中的类别（在本例中为 pos 和 neg）。仔细观察，你会发现这三个参数都是正则表达式。接下来的两行将验证是否正确加载语料库，并打印出语料库的相关类别和文件名。运行该程序，你会看到以下内容：

```
['austen-emma.txt', 'austen-persuasion.txt', 'austen-sense.txt',
'bible-kjv.txt',....]
[['The', 'Fulton', 'County', 'Grand', 'Jury', 'said', 'Friday',
'an', 'investigation', 'of',...]]
```

3. 现在已经确保了该语料库被正确加载，下面继续访问这两个类别中的任何一个示例文档。为此，我们首先创建一个列表，每个列表分别包含 pos 和 neg 两个类别的样本。添加以下两行代码：

```
posFiles = reader.fileids(categories='pos')
negFiles = reader.fileids(categories='neg')
```

reader.fileids() 方法的参数为类别名称。你可以发现，以上两行代码的目的是直接明了的。

4. 现在我们从 posFiles 和 negFiles 的列表中随机选择一个文件。为此，我们需要利用 Python random 库中的 randint() 函数。我们添加如下几行代码，接下来会详细说明它们的具体功能：

```
from random import randint
fileP = posFiles[randint(0,len(posFiles)-1)]
fileN = negFiles[randint(0, len(posFiles) - 1)]
print(fileP)
print(fileN)
```

第一行从 random 库中导入 randint() 函数。接下来分别从正面和负面类别评论集中随机选择一个文件。最后两行只是打印文件名。

5. 既然我们已经选择了两个文件，现在开始访问这两个文件并逐句打印在控制台上。使用与第一个实例相同的方法来逐行打印输出。添加以下代码行：

```
for w in reader.words(fileP):
  print(w + ' ', end='')
  if (w is '.'):
    print()
for w in reader.words(fileN):
  print(w + ' ', end='')
  if (w is '.'):
    print()
```

这些 for 循环逐一读取每个文件，并逐行打印在控制台上。完整的实例输出如下所示：

```
['neg', 'pos']
['neg/cv000_29416.txt', 'neg/cv001_19502.txt',
'neg/cv002_17424.txt', ...]
pos/cv182_7281.txt
neg/cv712_24217.txt
the saint was actually a little better than i expected it to be ,
in some ways .
in this theatrical remake of the television series the saint...
```

1.3.3　工作原理

这个实例的典型组成部分是 NLTK 中的 CategorizedPlaintextCorpusReader 类。由于下载的语料库已经分好类别，我们只需要在创建 reader 对象时提供适当的参数。实现 CategorizedPlaintextCorpusReader 类是从内部将样本加载到合适的位置（在该例中为 pos 和 neg）。

1.4　计算布朗语料库中三种不同类别的特殊疑问词

布朗语料库是 NLTK 数据包的一部分，是布朗大学最古老的文本语料库之一。它包含 500 个文本的集合，大致分为 15 个不同的类型 / 类别，如新闻、笑话、宗教等。这个语料库是纯文本分类语料库的典范，其中每个文本已经被分配了相应的主题 / 概念（有时是重叠的）。因此，你对它做的任何分析都与它对应的主题相一致。

1.4.1　准备工作

这个实例的目标是对任意一个给定的语料库执行一个简单的计数任务。在这个任务中将使用 NLTK 库的 FreqDist 对象。我们将在下一个实例中更详细地介绍 FreqDist，这里只关注它的应用。

1.4.2　如何实现

1. 创建一个名为 Brown WH.py 的新文件，首先添加以下导入语句：

```
import nltk
from nltk.corpus import brown
```

我们已经导入了 NLTK 库和布朗语料库。

2. 接下来，检查语料库中的所有类别，并从中任意选择三个类别来执行我们的任务：

```
print(brown.categories())
```

函数 brown.categories() 调用将返回布朗语料库中所有类别的列表。当运行这一行时，你会看到如下输出：

```
['adventure', 'belles_lettres', 'editorial', 'fiction',
'government', 'hobbies', 'humor', 'learned', 'lore', 'mystery',
'news', 'religion', 'reviews', 'romance', 'science_fiction']
```

3. 现在我们从这张列表中挑选出三种类别——小说（fiction）、笑话（humor）和爱情故事（romance），并从这三个类型文本中获取疑问词（whwords）用于后续计算：

```
genres = ['fiction', 'humor', 'romance']
whwords = ['what', 'which', 'how', 'why', 'when', 'where', 'who']
```

我们已经创建了一个包含 3 种所选类别的列表和一个包含 7 个疑问词的列表。

列表可长可短，取决于你认为哪些是疑问词。

4. 由于已生成需要计算的类别和单词列表，我们通常使用 for 循环来进行迭代并优化代码行数。首先我们在类别列表上编写一个 for 迭代器：

```
for i in range(0,len(genres)):genre = genres[i]
print()
print("Analysing '"+ genre + "' wh words")
genre_text = brown.words(categories = genre)
```

这四行代码只在类别列表上进行迭代，并将每个类别的整个文本作为连续列表词加载到 genre_text 变量中。

5. 接下来，我们将使用 NLTK 库的 FreqDist 对象进行一个复杂的简短声明。现在我们了解一下它的语法和广义输出结果：

```
fdist = nltk.FreqDist(genre_text)
```

FreqDist() 函数接收一个单词列表并返回一个对象，该对象包含映射词（map word）及其在输入单词列表中的相应频率。这里，fdist 对象将包含 genre_text 单词列表中每个词条（unique words）的频率。

6. 下一步我们将访问由 FreqDist() 函数返回的 fdist 对象，并得到每个疑问词的计数。

```
for wh in whwords:
print(wh + ':', fdist[wh], end=' ')
```

迭代疑问词列表，将每个疑问词作为索引访问 fdist 对象，获得所有这些疑问词的频率 / 计数，并将它们打印出来。运行完整的程序后会得到如下输出：

```
['adventure', 'belles_lettres', 'editorial', 'fiction',
'government', 'hobbies', 'humor', 'learned', 'lore', 'mystery',
'news', 'religion', 'reviews', 'romance', 'science_fiction']

Analysing 'fiction' wh words

what: 128 which: 123 how: 54 why: 18 when: 133 where: 76 who: 103

Analysing 'humor' wh words

what: 36 which: 62 how: 18 why: 9 when: 52 where: 15 who: 48

Analysing 'romance' wh words

what: 121 which: 104 how: 60 why: 34 when: 126 where: 54 who: 89
```

1.4.3 工作原理

在分析控制台的输出时，我们可以清楚地看到 3 种所选类别的所有 7 个疑问词的计数。通过计算疑问词的数量，我们可以在一定程度上衡量给定文本是否包含较多的关系从句或问句。同样，已知一个包含重要词汇的本体列表，我们可以通过词频计算来了解给定文本与该本体的相关性。词频计算和计数分布分析是所有文本分析中最古老、最简单、最流行的技巧之一。

1.5 探讨网络文本和聊天文本的词频分布

顾名思义，网络文本和聊天文本语料库是非正式的文献，它的内容来自于 Firefox 讨论论坛、电影脚本、酒类评论、个人广告和窃听对话的内容。在本节中我们的目标是掌握频率分布的使用及其特征 / 功能。

1.5.1 准备工作

为了完成这个实例的任务，我们将在 nltk.corpus.webtext 中的个人广告文件上运行频率分布计算。接下来，我们将探讨 nltk.FreqDist 对象的各种功能，如不同词的计数、最常用的 10 个词、最大频率词、频率分布图和制表。

1.5.2 如何实现

1. 创建一个名为 webtext.py 的新文件，并添加如下三行代码：

```
import nltk
from nltk.corpus import webtext
print(webtext.fileids())
```

我们已经导入了所需的库和 webtext 语料库。与此同时，我们还打印了组织文件的名称。运行该程序，你将看到如下输出：

```
['firefox.txt', 'grail.txt', 'overheard.txt', 'pirates.txt',
'singles.txt', 'wine.txt']
```

2. 选择包含个人广告数据的文件，并对该文件计算频率分布。添加如下三行代码：

```
fileid = 'singles.txt'
wbt_words = webtext.words(fileid)
fdist = nltk.FreqDist(wbt_words)
```

singles.txt 包含目标数据。因此，我们将该文件的单词加载到 wbt_words 中，并计算频率分布，获得 FreqDist 对象 fdist。

3. 添加以下几行代码，将显示出最常出现的单词（使用 fdist.max() 函数）和该单词的计数（使用 fdist [fdist.max()] 操作）：

```
print('Count of the maximum appearing token "',fdist.max(),'" : ',
fdist[fdist.max()])
```

4. 下面一行代码将使用 fdist.N() 函数得到频率分布包中不同单词的计数。在代码中添加如下行：

```
print('Total Number of distinct tokens in the bag : ', fdist.N())
```

5. 找出所选语料库中最常见的 10 个单词。函数 fdist.most_common() 可以实现该功能。在代码中添加以下两行：

```
print('Following are the most common 10 words in the bag')
print(fdist.most_common(10))
```

6. 使用 fdist.tabulate() 函数将整个频率分布制成表格。在代码中添加以下两行：

```
print('Frequency Distribution on Personal Advertisements')
print(fdist.tabulate())
```

7. 使用 fdist.plot() 函数绘制累积频率的频率分布图：

```
fdist.plot(cumulative=True)
```

运行程序得到输出：

```
['firefox.txt', 'grail.txt', 'overheard.txt', 'pirates.txt',
'singles.txt', 'wine.txt']

Count of the maximum appearing token " , " : 539

Total Number of distinct tokens in the bag : 4867

Following are the most common 10 words in the bag

[(',', 539), ('.', 353), ('/', 110), ('for', 99), ('and', 74),
('to',

4), ('lady', 68), ('-', 66), ('seeks', 60), ('a', 52)]
```

```
Frequency Distribution on Personal Advertisements
, . / for and to lady .........
539 353 110 99 74 74 .........
None
```

同时，弹出下图：

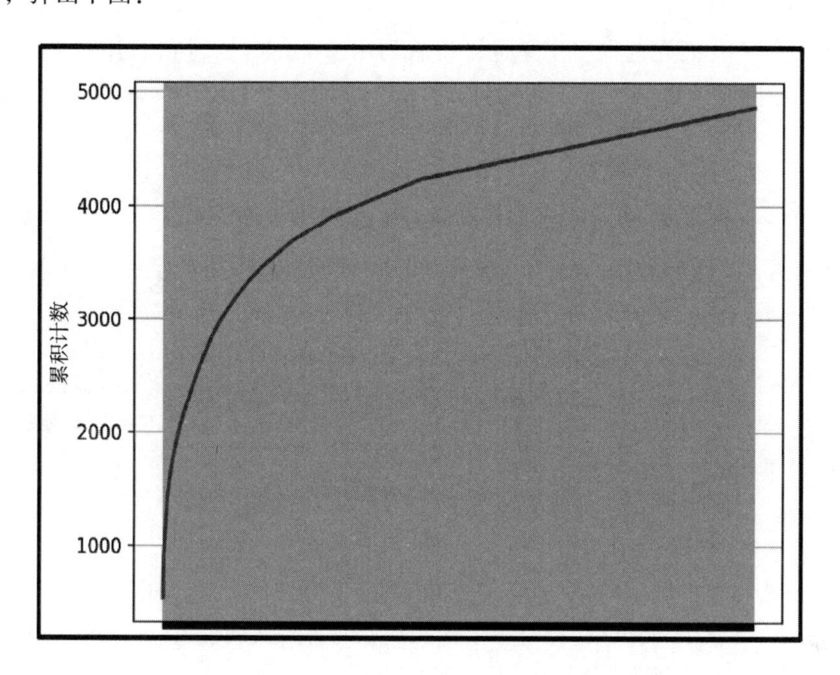

1.5.3 工作原理

通过分析输出结果，我们看到所有结果都很直观。但奇怪的是，大部分都是没有意义的，比如最大频率计数的词。当你查看 10 个最常见的词时，你也无法对目标数据集有太多的了解。原因在于没有对语料库进行预处理。在第 3 章中，我们将学习最基本的预处理步骤——停用词处理，到时候可以看到处理前后的差别。

1.6 使用 WordNet 进行词义消歧

从这个实例开始，我们将关注 WordNet。正如标题所示，我们将探讨单词的含义。总体上来说，英语是一种多歧义的语言，几乎所有的单词在不同的语境中都有不同的含义。举个最简单的单词，例如 bat，在地球上几乎任何地方的语言课程都将它作为前 10 个英语单词来进行学习。它的第一个意思是棍棒，在诸如板球、棒球、网球、壁球等各种运动中用于击球。

现在，bat 也有蝙蝠的含义，表示夜间飞行的哺乳动物。在 DC 漫画中，Bat 首选翻译成蝙蝠侠或者表示最先进的交通工具。这些都是名词变体。当 Bat 作为动词时，可以表示轻微的眨眼（眨动眼睑）。它还可以表示在一次打斗或竞赛中把某人打得血肉模糊。下面我们看具体实例。

1.6.1 准备工作

为达成目标，我们先选择一个单词，然后利用 WordNet 探讨它的各种含义。NLTK 已经配置了 WordNet，所以你不必安装任何其他库。我们选择另一个简单的单词 chair 作为本节的范例。

1.6.2 如何实现

1. 创建一个名为 ambiguity.py 的新文件，并添加如下代码行：

```
from nltk.corpus import wordnet as wn
chair = 'chair'
```

在这里导入所需的 NLTK corpus reader 类的 wordnet 作为 wn 对象。可以像使用其他 corpus readers 一样进行导入。然后，创建包含单词 chair 的字符串变量，为接下来的步骤做准备。

2. 这是最重要的一步。我们添加两行代码并进行详细解释：

```
chair_synsets = wn.synsets(chair)
print('Synsets/Senses of Chair :', chair_synsets, '\n\n')
```

第一行虽然看起来很简单，实际上是访问内部 WordNet 数据库的 API 接口，并获取与 chair 相关的所有含义。WordNet 把每一个含义都定义为同义词集（Synsets）。第二行只是打印出提取的内容。运行代码，将得到如下所示的输出：

```
Synsets/Senses of Chair : [Synset('chair.n.01'),
Synset('professorship.n.01'), Synset('president.n.04'),
Synset('electric_chair.n.01'), Synset('chair.n.05'),
Synset('chair.v.01'), Synset('moderate.v.01')]
```

正如你所看到的，列表中包含 7 个 Synset，这意味着在 WordNet 数据库中，单词 chair 存在 7 种不同的含义。

3. 添加以下 for 循环，它将迭代已经获得的 Synsets 列表并执行一些操作：

```
for synset in chair_synsets:
  print(synset, ': ')
  print('Definition: ', synset.definition())
  print('Lemmas/Synonymous words: ', synset.lemma_names())
  print('Example: ', synset.examples(), '\n')
```

我们已经迭代了 Synsets 列表，并打印出了每个含义的定义、相关的词条近义词以及每个含义在句子中的示例用法。一个典型的迭代将打印出如下内容：

```
Synset('chair.v.01') :
```

```
Definition: act or preside as chair, as of an academic department
in a university

Lemmas/Synonymous words: ['chair', 'chairman']

Example: ['She chaired the department for many years']
```

第一行是 Synset 的名字，第二行是 Synset 的定义，第三行包含与 Synset 相关的词条，第四行是一个例句。

我们将获得如下输出：

```
Synsets/Senses of Chair : [Synset('chair.n.01'),
Synset('professorship.n.01'), Synset('president.n.04'),
Synset('electric_chair.n.01'), Synset('chair.n.05'),
Synset('chair.v.01'), Synset('moderate.v.01')]

Synset('chair.n.01') :

Definition: a seat for one person, with a support for the back

Lemmas/Synonymous words: ['chair']

Example: ['he put his coat over the back of the chair and sat
down']

Synset('professorship.n.01') :

Definition: the position of professor

Lemmas/Synonymous words: ['professorship', 'chair']

Example: ['he was awarded an endowed chair in economics']

Synset('president.n.04') :

Definition: the officer who presides at the meetings of an
organization

Lemmas/Synonymous words: ['president', 'chairman', 'chairwoman',

chair', 'chairperson']

Example: ['address your remarks to the chairperson']

Synset('electric_chair.n.01') :

Definition: an instrument of execution by electrocution; resembles
an ordinary seat for one person

Lemmas/Synonymous words: ['electric_chair', 'chair', 'death_chair',
'hot_seat']

Example: ['the murderer was sentenced to die in the chair']

Synset('chair.n.05') :
```

```
Definition: a particular seat in an orchestra

Lemmas/Synonymous words: ['chair']

Example: ['he is second chair violin']

Synset('chair.v.01') :

Definition: act or preside as chair, as of an academic department
in a university

Lemmas/Synonymous words: ['chair', 'chairman']

Example: ['She chaired the department for many years']

Synset('moderate.v.01') :

Definition: preside over

Lemmas/Synonymous words: ['moderate', 'chair', 'lead']

Example: ['John moderated the discussion']
```

1.6.3　工作原理

在输出中，你可以看到单词 chair 包含的 7 个含义的定义、词条和例句。前面的代码示例详细说明了每个操作都有直接的 API 接口。现在我们来谈谈 WordNet 如何得出这样的结论。WordNet 是一个单词数据库，以分层的方式存储单词的所有信息。关于同义词集和 WordNet 的分层存储特性，下一节将进行更详细的介绍。

1.7　选择两个不同的同义词集，使用 WordNet 探讨上位词和下位词的概念

在上一节我们已经介绍了下位词（hyponym），与通用词（比如 bat）相比较具有更具体的含义。更具体的含义指板球棒、棒球棒、食肉蝙蝠、壁球拍，等等。这些下位词使我们在沟通中想要表达的含义更加具体。

与下位词相反，上位词（hypernym）是更具有一般性的形式或词，包含相同的概念。就之前的例子而言，bat 是一个更通用的词，它可能表示棍棒、棒、手工品、哺乳动物、动物或生物体。我们可以继续将它泛化成物质实体、生物或物体来作为 bat 这个词的上位词。

1.7.1　准备工作

为了探讨上位词和下位词的概念，我们选择同义词集 bed.n.01（bed 的第一个词义）和 woman.n.02（woman 的第二个词义）作为范例。在本节我们将介绍上位词和下位词 API 的用法和含义。

1.7.2 如何实现

1. 创建一个名为 HypoNHypernyms.py 的新文件，并添加以下三行代码：

```
from nltk.corpus import wordnet as wn
woman = wn.synset(woman.n.02')
bed = wn.synset('bed.n.01')
```

导入这些库，并初始化这两个同义词集以供后续使用。

2. 添加以下两行代码：

```
print(woman.hypernyms())
woman_paths = woman.hypernym_paths()
```

这是 hypernyms() API 函数在 woman 同义词集上的简单调用，它将返回具有直系关系的同义词集。但是，hypernym_paths() 函数有点棘手，它将返回一个集合列表。每个集合包含从根结点到 woman 同义词集的路径。当运行这两句时，我们在控制台中可以看到 woman 同义词集的两个直系关系：

```
[Synset('adult.n.01'), Synset('female.n.02')]
```

woman 隶属于 WordNet 数据库层次结构中的成人和女性类别。

3. 现在我们尝试打印从根结点到 woman.n.01 结点的路径。为此，请添加以下代码行并嵌入 for 循环：

```
for idx, path in enumerate(woman_paths):
  print('\n\nHypernym Path :', idx + 1)
for synset in path:
  print(synset.name(), ', ', end='')
```

如上所述，返回的对象是遵循从根结点到 woman.n.01 结点的路径进行排列的集合列表，并存储在 WordNet 的层次结构中。当运行时，出现如下路径示例 Path：

```
Hypernym Path : 1

entity.n.01 , physical_entity.n.01 , causal_agent.n.01 ,
person.n.01 , adult.n.01 , woman.n.01
```

4. 现在开始我们研究下位词。以下两行将获取同义词集 bed.n.01 的下位词并将其打印到控制台：

```
types_of_beds = bed.hyponyms()
print('\n\nTypes of beds(Hyponyms): ', types_of_beds)
```

如上所述，运行代码，你将得到以下 20 个同义词集：

```
Types of beds(Hyponyms): [Synset('berth.n.03'), Synset('built-
in_bed.n.01'), Synset('bunk.n.03'), Synset('bunk_bed.n.01'),
Synset('cot.n.03'), Synset('couch.n.03'), Synset('deathbed.n.02'),
Synset('double_bed.n.01'), Synset('four-poster.n.01'),
Synset('hammock.n.02'), Synset('marriage_bed.n.01'),
Synset('murphy_bed.n.01'), Synset('plank-bed.n.01'),
Synset('platform_bed.n.01'), Synset('sickbed.n.01'),
```

```
Synset('single_bed.n.01'), Synset('sleigh_bed.n.01'),
Synset('trundle_bed.n.01'), Synset('twin_bed.n.01'),
Synset('water_bed.n.01')]
```

在 WordNet 中，这些是单词词义 bed.n.01 的下位词或更具体的术语。

5. 现在我们添加以下代码行，打印出对人类来说更有意义的实际单词或词条（lemma）：

```
print(sorted(set(lemma.name() for synset in types_of_beds for lemma
in synset.lemmas())))
```

这行代码非常类似于之前上位词示例中我们编写的四行代码嵌套了 for 循环所做的操作，而这里是一行操作（换句话说，这里展示了 Python 的魅力）。这行代码将打印出 26 条非常有意义且具有特殊含义的词条。我们看看最后的输出：

```
Output: [Synset('adult.n.01'), Synset('female.n.02')]
Hypernym Path : 1
entity.n.01 , physical_entity.n.01 , causal_agent.n.01 ,
person.n.01 , adult.n.01 , woman.n.01 ,
Hypernym Path : 2
entity.n.01 , physical_entity.n.01 , object.n.01 , whole.n.02 ,
living_thing.n.01 , organism.n.01 , person.n.01 , adult.n.01 ,
woman.n.01 ,
Hypernym Path : 3
entity.n.01 , physical_entity.n.01 , causal_agent.n.01 ,
person.n.01 , female.n.02 , woman.n.01 ,
Hypernym Path : 4
entity.n.01 , physical_entity.n.01 , object.n.01 , whole.n.02 ,
living_thing.n.01 , organism.n.01 , person.n.01 , female.n.02 ,
woman.n.01 ,

Types of beds(Hyponyms): [Synset('berth.n.03'), Synset('built-
in_bed.n.01'), Synset('bunk.n.03'), Synset('bunk_bed.n.01'),
Synset('cot.n.03'), Synset('couch.n.03'), Synset('deathbed.n.02'),
Synset('double_bed.n.01'), Synset('four-poster.n.01'),
Synset('hammock.n.02'), Synset('marriage_bed.n.01'),
Synset('murphy_bed.n.01'), Synset('plank-bed.n.01'),
Synset('platform_bed.n.01'), Synset('sickbed.n.01'),
Synset('single_bed.n.01'), Synset('sleigh_bed.n.01'),
Synset('trundle_bed.n.01'), Synset('twin_bed.n.01'),
Synset('water_bed.n.01')]

['Murphy_bed', 'berth', 'built-in_bed', 'built_in_bed', 'bunk',
'bunk_bed', 'camp_bed', 'cot', 'couch', 'deathbed', 'double_bed',
'four-poster', 'hammock', 'marriage_bed', 'plank-bed',
'platform_bed', 'sack', 'sickbed', 'single_bed', 'sleigh_bed',
'truckle', 'truckle_bed', 'trundle', 'trundle_bed', 'twin_bed',
'water_bed']
```

1.7.3　工作原理

如上所述，woman.n.01 有 2 个上位词，即成人和女性。但是，在输出中显示的是在 WordNet 数据库层次结构中从实体根结点到 woman 的四条不同路径。

同样，bed.n.01 同义词集有 20 个下位词，它们的含义更加具体且明确。由于几乎没有歧义，在层次结构中，下位词通常与叶子结点或非常接近叶子的结点相对应。

1.8　基于 WordNet 计算名词、动词、形容词和副词的平均多义性

首先我们来了解什么是多义性。多义性意味着一个单词或短语有许多种可能的含义。正如之前看到的那样，英语是一种多歧义的语言，对层级结构中的大多数单词来说，通常包含一种以上的含义。我们必须根据 WordNet 中所有单词特定的语言特性来计算平均多义性。不同于前面的实例，本节实例不仅会涉及 API 的概念，还会涉及语言概念。

1.8.1　准备工作

我们会展示如何计算一个单词在某种词性（POS）下的多义性，对于其余 3 类词性的多义性计算，你只需要修改相应程序即可。我们会在实例中提供足够的提示，让你更容易掌握。下面我们来看看单独计算名词的平均多义性的实例。

1.8.2　如何实现

1. 创建一个名为 polysemy.py 的新文件，并添加两行初始化代码：

```
from nltk.corpus import wordnet as wn
type = 'n'
```

我们对感兴趣的词性进行初始化。当然，我们还导入了所需的库。为了描述方便，我们用 n 表示名词。

2. 该实例最重要的代码行如下：

```
synsets = wn.all_synsets(type)
```

这个 API 返回 WordNet 数据库中名词类型的所有同义词集。同样，如果将词性改为动词、副词或形容词，API 将返回相应类型的所有单词（提示 # 1）。

3. 把每个同义词集的所有词条合并成一个可以进一步处理的大列表。添加如下代码：

```
lemmas = []
for synset in synsets:
  for lemma in synset.lemmas():
    lemmas.append(lemma.name())
```

这段代码非常直观，用一个嵌套的 for 循环迭代同义词集（synsets）列表和每个同义词集（synset）中的词条（lemmas）列表，并将它们添加到词条（lemmas）大列表中。

4. 虽然这个大列表包含了所有词条，但仍存在一个问题。它是一个列表，所以有一些重复的内容，我们需要删除重复项：

```
lemmas = set(lemmas)
```

这里将一个列表转换成一个集合，就可以自动删除其中重复的内容。

5. 该实例第二重要的步骤如下。计算 WordNet 数据库中每个词条的含义：

```
count = 0
for lemma in lemmas:
  count = count + len(wn.synsets(lemma, type))
```

大部分代码很易懂，我们把重点放在 wn.synsets API（lemma，type）上。这个 API 的输入是一个单词 / 词条（作为第一个参数）及其所属的词性，返回该单词 / 词条的所有含义。需要注意的是，输出取决于你提供的词性类别，因为它仅返回给定词性下的所有含义（提示 # 2）。

6. 现在已有平均多义性计算所需的所有数值，接下来只需要计算并打印出结果：

```
print('Total distinct lemmas: ', len(lemmas))
print('Total senses :',count)
print('Average Polysemy of ', type,': ' , count/len(lemmas))
```

这里输出了所有不同词条的数目，所有含义的数目以及名词词性的平均多义性数目：

```
Output: Total distinct lemmas: 119034
Total senses : 152763
Average Polysemy of n : 1.2833560159282222
```

1.8.3 工作原理

我们介绍一下关于如何计算其他单词词性的平均多义性。正如实例所示，名词表示为 n。类似地，动词表示为 v，副词表示为 r，形容词表示为 a（提示 #3）。

现在，我们已经给了你足够的提示，继续编写 NLP 程序吧。

CHAPTER 2

第2章

针对原始文本，获取源数据和规范化

2.1 引言

在前一章中，我们介绍了 NLTK 内置的语料库。通常情况下，语料库组织良好，使用规范，但是当我们解决实际工程问题时，情况并不总是如此。我们可能获取不到统一格式的数据，更不用说获取到规范化和结构化的数据。本章的目标是介绍一些 Python 库，帮助你从类似于 PDF 和 Word DOCX 这样的二进制文件中提取数据。我们也将了解和学习如何从网络信息源（web feeds）（如 RSS）中获取数据，以及利用一个库帮助解析 HTML 文本并从文档中提取原始文本。我们还将学习如何从不同来源提取原始文本，对其进行规范化，并基于它创建一个用户定义的语料库。

在本章中，你将学习 7 个不同的实例。正如本章题目所示，我们将学习从 PDF 文件、Word 文档和 Web 中获取数据。PDF 和 Word 文档是二进制文件，通过 Web，你将获得 HTML 格式的数据，因此，我们也会对数据执行规范化和原始文本转换任务。

2.2 字符串操作的重要性

作为一名 NLP 专家，你将要处理大量的文本内容。当你在处理文本时，你必须知道一些字符串操作。我们将从几个简短的范例入手，帮助你理解 str 类及其在 Python 中的相关操作。

2.2.1 准备工作

在本节中，你仅仅需要 Python 解释器和一个文本编辑器。我们将使用 join（连接）、split（分割）、addition（加法）和 multiplication（乘法）运算符以及索引。

2.2.2 如何实现

1. 创建一个新的 Python 文件，命名为 StringOps1.py。

2. 定义以下两个对象：

```
namesList = ['Tuffy','Ali','Nysha','Tim' ]
sentence = 'My dog sleeps on sofa'
```

第一个对象 nameList 是一个包含若干名字的字符串列表，第二个对象 sentence 是一个包含一句话的字符串对象。

3. 首先，我们看看 join 函数的特点以及它的功能：

```
names = ';'.join(namesList)
print(type(names), ':', names)
```

join() 函数可以被任意一个 string 对象调用，它的输入参数是一个 str 对象的列表。通过将调用字符串的内容作为连接分隔符，它将所有 str 对象连接成一个 str 对象，并返回连接后的对象。运行这两行代码后，你得到的输出如下：

```
<class 'str'> : Tuffy;Ali;Nysha;Tim
```

4. 接下来，我们来看 split 方法的功能：

```
wordList = sentence.split(' ')
print((type(wordList)), ':', wordList)
```

当 split 函数调用一个字符串时，它会将其内容分割为多个 str 对象，创建一个包含这些字符串对象的列表，并返回该列表。该函数接受单个 str 对象作为参数，表示分隔符。运行代码，得到如下输出：

```
<class 'list'> : ['My', 'dog', 'sleeps', 'on', 'sofa']
```

5. 算术运算符 + 和 * 也可以用于字符串。添加以下代码并输出：

```
additionExample = 'ganehsa' + 'ganesha' + 'ganesha'
multiplicationExample = 'ganesha' * 2
print('Text Additions :', additionExample)
print('Text Multiplication :', multiplicationExample)
```

我们首先看一下输出结果，随后讨论其工作原理：

```
Text Additions: ganehsaganeshaganesha
Text Multiplication: ganeshaganesha
```

+ 运算符被称为连接符，它将字符串连接为单个 str 对象，产生一个新的字符串。如前所述，我们也可以使用 * 运算符对字符串做乘法。此外，需要注意的是这些操作不会添加任何额外的内容，例如在字符串之间插入空格。

6. 接下来，我们来了解一下字符串中的字符索引。添加下列几行代码：

```
str = 'Python NLTK'
print(str[1])
print(str[-3])
```

首先，我们声明一个新的 string 对象。然后可以直接访问字符串中的第二个字符（y）。这里还有个小技巧：Python 允许你在访问任何列表对象时使用负索引，比如说 -1 意味着最后一个成员，-2 是倒数第二个成员，依此类推。例如，在前面代码的 str 对象中，索引 7 和 -4 是相同的字符 N：

```
Output: <class 'str'> : Tuffy;Ali;Nysha;Tim
<class 'list'> : ['My', 'dog', 'sleeps', 'on', 'sofa']
Text Additions : ganehsaganeshaganesha
Text Multiplication : ganeshaganesha
y
L
```

2.2.3 工作原理

我们使用 split() 函数将一个字符串变成了一个字符串列表，并使用 join() 函数将一个字符串列表变成了一个字符串。接下来我们了解了有关字符串的一些算术运算符的用法。需要注意的是，我们不能在字符串中使用“-”（负号）和“/”（除法）运算符。最后，我们了解了如何在任一字符串中访问单个字符，特别值得一提的是，我们可以在访问字符串时使用负索引。

本节实例非常简单和直观，主要是介绍 Python 允许的一些常见和不常见的字符串操作。接下来，我们将在以上操作基础上继续学习一些字符串操作。

2.3 深入实践字符串操作

接下来，我们将了解子字符串、字符串替换以及如何访问一个字符串的所有字符。

2.3.1 如何实现

1. 创建一个新的 Python 文件，命名为 StringOps2.py 并定义以下 string 对象：

```
str = 'NLTK Dolly Python'
```

2. 访问 str 对象中以第四个字符作为结束的子串。

```
print('Substring ends at:',str[:4])
```

我们知道索引从零开始，因此将返回由第 0 个到第 3 个字符组成的子串。运行代码，输出如下：

```
Substring ends at: NLTK
```

3. 访问 str 对象中从某个点开始直到末尾的子串：

```
print('Substring starts from:',str[11:] )
```

以上代码指示解释器返回 str 对象中从索引 11 到结束的一个子串。运行代码，得到以下输出：

```
Substring starts from: Python
```

4. 从 str 对象中访问包含 Dolly 的子串。添加以下行：

```
print('Substring :',str[5:10])
```

以上代码返回从索引 5 到 10 的字符，不包括第 10 个字符。输出是：

```
Substring : Dolly
```

5. 我们在前一节中已经了解了负索引在字符串操作中的应用。现在我们试试看它在获取子串中的作用：

```
print('Substring fancy:', str[-12:-7])
Run and check the output, it will be -
Substring fancy: Dolly
```

这里得到的输出与上一步完全相同！为了理解这个结果，我们做一些计算：-1 表示最后一个字符，-2 是倒数第二个字符，依此类推。你将会发现 [5:10] 和 [-12:-7] 在这个例子中得出的子串是相同的。

6. 了解 in 操作符在 if 语句中的用法：

```
if 'NLTK' in str:
  print('found NLTK')
```

运行以上代码，程序的输出如下所示：

```
found NLTK
```

如上所示，in 操作符会检查左边的字符串是否属于右边字符串的子串。

7. 了解 str 对象中 replace 函数的使用：

```
replaced = str.replace('Dolly', 'Dorothy')
print('Replaced String:', replaced)
```

replace 函数只需要两个参数。第一个是需要被替换的子字符串，第二个是用来替换前面子字符串的新子字符串。replace 函数返回一个新的 string 对象，并且它不会修改调用它的字符串，运行代码，有如下输出：

```
Replaced String: NLTK Dorothy Python
```

8. 最后，迭代上面得到的 replaced 对象并访问它的每一个字符：

```
print('Accessing each character:')
for s in replaced:
  print(s)
```

以上操作每次在新的一行输出 replaced 对象的每个字符，最终输出如下：

```
Output: Substring ends at: NLTK
Substring starts from: Python
Substring : Dolly
```

```
Substring fancy: Dolly
found NLTK
Replaced String: NLTK Dorothy Python
Accessing each character:
N
L
T
K
D
o
r
o
t
h
y
P
y
t
h
o
n
```

2.3.2 工作原理

字符串对象只是一个字符列表。正如第一步所示，我们可以像访问一个列表那样用 for 语句来访问字符串中的每个字符。任何列表的方括号内的字符"："表示我们想要的一个子列表。方括号内，如果字符"："之后是一个数字 n，表示我们希望获得一个从列表索引 0 开始到索引 n-1 结束的子列表。同样地，一个数字 m 后跟着字符"："，则表示我们想要一个从列表索引 m 开始到列表末尾的子列表。

2.4 在 Python 中读取 PDF 文件

本节第一个实例是从 Python 中访问 PDF 文件。首先，你需要安装 PyPDF2 库。

2.4.1 准备工作

假设你已经安装了 pip。然后，在 Python2 或 Python3 版本上用 pip 安装 PyPDF2 库，你只需要在命令行中运行以下命令：

```
pip install pypdf2
```

如果你成功安装了 PyPDF2 库，就完成了准备工作。与此同时，你需要通过以下链接下载一些我们将在本节用到的测试文档：https://www.dropbox.com/sh/bk18dizhsu1p534/AABEuJw4TArUbzJf4Aa8gp5Wa?dl=0.

2.4.2 如何实现

1. 创建一个新的 Python 文件，命名为 pdf.py 并添加以下代码：

```
from PyPDF2 import PdfFileReader
```

这行代码会导入 PyPDF2 库中的 PdfFileReader 类。

2. 在上面创建的文件中添加如下 Python 函数，它的功能是读取一个 PDF 文件并返回其全文：

```
def getTextPDF(pdfFileName, password = '')
```

该函数需要两个参数，一个是你要读取的 PDF 文件路径，一个是这个 PDF 文件的密码（如果有的话）。可见，password 参数是可选的。

3. 现在我们来定义这个函数。在该函数下添加如下代码：

```
pdf_file = open(pdfFileName, 'rb')
read_pdf = PdfFileReader(pdf_file)
```

第一行代码会以读取和反向查找模式打开文件。第一行本质是一个 Python 文件打开命令/函数，仅能打开非文本的二进制文件。第二行将打开的文件传递给 PdfFileReader 类，用于处理 PDF 文档。

4. 如果文件设置了密码保护，接下来是解密被密码保护的 PDF 文件：

```
if password != '':
  read_pdf.decrypt(password)
```

如果在函数调用时设置了密码，那么我们在解密这个文件时也同样需要密码。

5. 从 PDF 文件中读取文本：

```
text = []
for i in range(0,read_pdf.getNumPages()-1):
  text.append(read_pdf.getPage(i).extractText())
```

创建一个字符串列表，并将每一页的文本都添加到这个列表中。

6. 返回最终的输出结果：

```
return '\n'.join(text)
```

将列表中所有的字符串都连接起来，并且在每个字符串之间都加一个换行符，返回连接后的单一字符串。

7. 在 pdf.py 目录下创建另一个名为 TestPDFs.py 的文件，添加以下导入语句：

```
import pdf
```

8. 现在我们打印输出两个文档中的文本，其中一个是受密码保护的，一个是未加密的：

```
pdfFile = 'sample-one-line.pdf'
pdfFileEncrypted = 'sample-one-line.protected.pdf'
print('PDF 1: \n',pdf.getTextPDF(pdfFile))
print('PDF 2: \n',pdf.getTextPDF(pdfFileEncrypted,'tuffy'))
```

输出：本实例的前六步只是创建了一个 Python 函数，并不向控制台输出内容，第七和第八步会输出以下内容：

```
This is a sample PDF document I am using to demonstrate in the
tutorial.

This is a sample PDF document

password protected.
```

2.4.3　工作原理

PyPDF2 是用于提取 PDF 文件内容的一个纯 Python 库。该库有很多功能，可用于裁剪页面、叠加图像数字签名、创建新的 PDF 文件等。但是，对 NLP 工程师需要实现的文本分析任务来说，该库只用来读取内容。在第二步中，以反向查找模式打开文件很重要，因为当加载文件内容时，PyPDF2 模块试图从尾部开始读取文件内容。此外，如果 PDF 文件是受密码保护的，而你没有在访问文件前解密文件，Python 解释器将抛出一个 PdfReadError 错误。

2.5　在 Python 中读取 Word 文件

在该节中，我们将学习如何加载和读取 Word/DOCX 文档。用于读取 Word/DOCX 文件的相关库会更加全面，在这些库中我们还可以处理段落边界、文本样式以及对所谓的 run 对象的操作。我们将会了解以上提到的所有内容，因为这些内容在文本分析任务中是至关重要的。

TIP　如果你没有安装 Microsoft Word 软件，你可以使用 Liber Office 和 Open Office 软件的开源版本来创建和编辑“.docx”文件。

2.5.1　准备工作

假设你已经在你的机器上安装了 pip，我们将使用 pip 来安装 python-docx 库。不要将它与另一个名为 docx 的库混淆，这是两个完全不同的库。我们将从 python docx 库中导入 docx 对象。在命令行中执行下面的命令将安装这个库：

```
pip install python-docx
```

成功安装了该库后，继续下一步，我们将在这个实例中使用一个测试文档，如果你已经通过本章第一小节提供的链接下载了所有文档，你应该已具备相关文档。如果没有，请从 https://www.dropbox.com/sh/bk18dizhsu1p534/AABEuJw4TArUbzJf4Aa8gp5Wa?dl=0 下载 sample-one-line.docx 文档。

现在，准备工作就全部完成了。

2.5.2　如何实现

1. 创建一个新的 Python 文件，命名为 word.py 并添加以下导入代码：

```
import docx
```

这里只需导入 python-docx 模块的 docx 对象。

2. 定义 getTextWord 函数：

```
def getTextWord(wordFileName):
```

该函数需要一个字符串参数 wordFileName，包含你要读取的 Word 文件的绝对路径。

3. 初始化 doc 对象：

```
doc = docx.Document(wordFileName)
```

此时 doc 对象加载了你要读取的 Word 文件。

4. 接下来我们要从已经加载文档的 doc 对象中读取文本，添加以下代码来实现：

```
fullText = []
for para in doc.paragraphs:
  fullText.append(para.text)
```

首先初始化一个字符串列表 fullText，然后采用 for 循环逐段从文档中读取文本，并把每段都放到 fullText 列表中去。

5. 然后，我们将所有的片段 / 段落连接为一个字符串对象，并将其作为函数的输出结果返回：

```
return '\n'.join(fullText)
```

通过以上操作，我们将 fullText 数组的所有元素用“\ n”分隔符连接起来，并返回连接后的对象。最后保存该 Python 文件并退出。

6. 创建另一个 Python 文件，命名为 TestDocX.py，并添加以下导入声明：

```
import docx
import word
```

这里只需导入 docx 库以及我们在前五步中实现的 word.py 文件。

7. 现在我们将要读取一个 DOCX 文件并使用我们在 word.py 中实现的 API 打印输出它的全部内容。添加以下两行代码：

```
docName = 'sample-one-line.docx'
print('Document in full :\n',word.getTextWord(docName))
```

首先在第一行代码中初始化文档的路径，然后使用 API 打印输出文档的全部内容。当你运行这部分代码时，得到以下输出：

```
Document in full :
```

这是一个带有一些粗体文本、一些斜体文本和一些下划线文本的 PDF 示例文档。我们还嵌入了一个标题，如下所示：

```
This is my TITLE.
This is my third paragraph.
```

8. 正如前面提到的，Word / DOCX 文档是一个更加丰富的信息来源，除了提供文本内容外，还能提供很多信息。现在我们来看有关段落的信息。添加以下四行代码：

```
doc = docx.Document(docName)
print('Number of paragraphs :',len(doc.paragraphs))
print('Paragraph 2:',doc.paragraphs[1].text)
print('Paragraph 2 style:',doc.paragraphs[1].style)
```

以上代码的第二行打印出了给定文档中段落的数量。第三行打印出了文档中第二段的内容。而第四行将会打印出第二段的样式，比如在这个例子中的样式就是 Title 类型。当你运行以上代码后，输出将如下所示：

```
Number of paragraphs : 3
Paragraph 2: This is my TITLE.
Paragraph 2 style: _ParagraphStyle('Title') id: 4374023248
```

9. 接下来，我们将了解什么是 run 对象。添加以下代码：

```
print('Paragraph 1:',doc.paragraphs[0].text)
print('Number of runs in paragraph 1:',len(doc.paragraphs[0].runs))
for idx, run in enumerate(doc.paragraphs[0].runs):
  print('Run %s : %s' %(idx,run.text))
```

首先，我们获得文档第一段的全部内容。然后，我们获得第一段中 run 对象的数目。最后，我们把每个 run 对象打印输出。

10. 为了明确每个 run 对象的格式，添加以下代码：

```
print('is Run 0 underlined:',doc.paragraphs[0].runs[5].underline)
print('is Run 2 bold:',doc.paragraphs[0].runs[1].bold)
print('is Run 7 italic:',doc.paragraphs[0].runs[3].italic)
```

这段代码的各行分别在检查相应 run 对象的下划线样式、粗体样式以及斜体样式。最终输出如下：

```
Output: Document in full :
This is a sample PDF document with some text in BOLD, some in
ITALIC and some underlined. We are also embedding a Title down
below.
This is my TITLE.
This is my third paragraph.
Number of paragraphs : 3
Paragraph 2: This is my TITLE.
Paragraph 2 style: _ParagraphStyle('Title') id: 4374023248
Paragraph 1: This is a sample PDF document with some text in BOLD,
some in ITALIC and some underlined. We're also embedding a Title
down below.
Number of runs in paragraph 1: 8
Run 0 : This is a sample PDF document with
Run 1 : some text in BOLD
Run 2 : ,
```

```
Run 3 : some in ITALIC
Run 4 :  and
Run 5 : some underlined.
Run 6 :  We are also embedding a Title down below
Run 7 : .
is Run 0 underlined: True
is Run 2 bold: True
is Run 7 italic: True
```

2.5.3 工作原理

首先，我们在 word.py 文件中写了一个函数，它将读取给定的 DOCX 文件并返回一个包含文件全部内容的字符串对象。前面的输出内容大都是不需要解释的，我特别阐述了关于 Paragraph 和 Run 的输出内容。DOCX 文件的结构可以用 python-docx 库的三个数据类型来表示，其中最高一级是 Document 对象。

每个文档都包含多个段落。文档中出现新的一行或一个回车，就表示开始一个新的段落。每个段落用多个 Run 对象表示段落内格式的变化，这里的格式包含有字体、尺寸、颜色和其他样式元素（如粗体、斜体、下划线等等）。这些元素每次发生变化时，都会创建一个新的 Run 对象。

2.6 使用 PDF、DOCX 和纯文本文件，创建用户自定义的语料库

本节不包含新的库或者概念。我们将重新用到第一章中的语料库的概念，只不过现在我们要创建自己的语料库，而不是使用从互联网上得到的语料库。

2.6.1 准备工作

在准备方面，我们将使用本章第一个实例中提到的 Dropbox 文件夹中的几个文件。如果你已经从那个文件夹中下载了全部的文件，那么你已经完成了准备工作。否则，请从 https://www.dropbox.com/sh/bk18dizhsu1p534/AABEuJw4TArUbzJf4Aa8gp5Wa?dl=0 下载如下文件：

- sample_feed.txt
- sample-pdf.pdf
- sample-one-line.docx

如果你没有按照本章的顺序来完成实例，那么你需要先回头看看本章的前两个实例。我们将用到本章前两个实例中完成的两个模块 word.py 和 pdf.py。本节实例更多是关于本章前两个实例所做工作的应用以及第一章中关于语料库概念的应用。下面我们来看实际的代码。

2.6.2　如何实现

1. 创建一个新的 Python 文件，命名为 createCorpus.py 并添加以下代码：

```
import os
import word, pdf
from nltk.corpus.reader.plaintext import PlaintextCorpusReader
```

我们导入 os 库用于与文件有关的操作，word 库和 pdf 库是本章前两节完成的库，最后导入的 PlaintextCorpusReader 是为了完成语料库建立这一最终目标。

2. 编写一个简单的函数，用来打开并读取一个纯文本文件，并将其全部内容作为 string 对象返回。添加以下代码：

```
def getText(txtFileName):
  file = open(txtFileName, 'r')
  return file.read()
```

第一行代码定义了函数及其输入参数。第二行代码以只读方式打开文件（open 函数的第二个参数 r 表示以只读方式打开）。第三行代码读取打开文件的内容并将其作为 string 对象返回。

3. 在磁盘或文件系统中创建一个新文件夹 corpus。添加以下三行代码：

```
newCorpusDir = 'mycorpus/'
if not os.path.isdir(newCorpusDir):
  os.mkdir(newCorpusDir)
```

第一行定义的 string 对象包含了新文件夹名，第二行检查该文件夹在磁盘或文件系统中是否存在，第三行则通过执行 os.mkdir() 函数在磁盘上创建一个给定名字的文件夹。以上代码执行后将在你的 Python 文件所在的工作目录下创建一个名为 mycorpus 的新文件夹。

4. 然后，逐个读取前面提到的三个文件。首先从纯文本文件开始，添加以下代码：

```
txt1 = getText('sample_feed.txt')
```

调用之前完成的 getText 函数，它将读取 Sample_feed.txt 文件并将输出结果存入名为 txt1 的字符串对象中。

5. 现在，添加以下代码来读取 PDF 文件：

```
txt2 = pdf.getTextPDF('sample-pdf.pdf')
```

这里使用了 PDF.py 模块的 getTextPDF() 函数，它将读取 sample-pdf.pdf 文件并将文件内容存入名为 txt2 的字符串对象中。

6. 最后，通过以下代码读取 DOCX 文件：

```
txt3 = word.getTextWord('sample-one-line.docx')
```

这里使用了 word.py 模块的 getTexWord() 函数，它将读取 sample-one-line.docx 文件并将文件内容存入名为 txt3 的字符串对象中。

7. 接下来，将上面读到的三个字符串对象写到磁盘文件中。添加以下代码：

```
files = [txt1,txt2,txt3]
for idx, f in enumerate(files):
  with open(newCorpusDir+str(idx)+'.txt', 'w') as fout:
    fout.write(f)
```

- **第一行**：创建一个包含以上三个字符串对象的数组
- **第二行**：使用 for 循环来遍历 files 数组
- **第三行**：以只写模式打开一个新文件（采用 w 选项调用 open 函数）
- **第四行**：将当前字符串内容写到文件中

8. 在 mycorpus 目录下，也就是我们之前存放文件的目录下新建一个 PlainTextCorpus 对象：

```
newCorpus = PlaintextCorpusReader(newCorpusDir, '.*')
```

以上一行代码看似简单，但是它在内部做了很多的文本处理，如识别段落、句子、单词等等。该函数的两个参数分别是语料库目录的路径以及要处理的文件名模式（这里我们已经设置 corpus reader 可以处理该目录下所有的文件）。通过以上步骤，我们创建了一个用户自定义的语料库。

9. 接下来，我们来看 PlainTextCorpusReader 是否加载正常。添加以下代码来进行测试：

```
print(newCorpus.words())
print(newCorpus.sents(newCorpus.fileids()[1]))
print(newCorpus.paras(newCorpus.fileids()[0]))
```

第一行代码将打印输出语料库包含的所有单词数组（部分）。第二行代码将打印输出文件 1.txt 中的句子。第三行代码将打印输出文件 0.txt 中的段落：

```
Output: ['Five', 'months', '.', 'That', "'", 's', 'how', ...]
[['A', 'generic', 'NLP'], ['(', 'Natural', 'Language',
'Processing', ')', 'toolset'], ...]
[[['Five', 'months', '.']], [['That', "'", 's', 'how', 'long',
'it', "'", 's', 'been', 'since', 'Mass', 'Effect', ':',
'Andromeda', 'launched', ',', 'and', 'that', "'", 's', 'how',
'long', 'it', 'took', 'BioWare', 'Montreal', 'to', 'admit', 'that',
'nothing', 'more', 'can', 'be', 'done', 'with', 'the', 'ailing',
'game', "'", 's', 'story', 'mode', '.'], ['Technically', ',', 'it',
'wasn', "'", 't', 'even', 'a', 'full', 'five', 'months', ',', 'as',
'Andromeda', 'launched', 'on', 'March', '21', '.']], ...]
```

2.6.3　工作原理

该实例最后一步的输出很简单直接，展示了各个对象不同的特征。输出内容的第一行是新语料库的单词列表，它与句子、段落、文件等更高级的结构没有关系。第二行是 1.txt 文件中所有句子组成的列表，其中每个句子都是由该句子中单词组成的列表。第三行是 0.txt 文件中所有段落组成的列表，其中每个段落对象又是由该段落中的句子组成的列表。从中可以发现，这些段落和句子保留了很多原有的结构。

2.7 读取 RSS 信息源的内容

丰富网站摘要（Rich Site Summary，RSS）信息源（feed）是一种计算机可读格式，用于传送互联网上定期更新的内容。大多数提供通知信息的网站以这种格式提供更新，例如新闻文章、在线出版物等。订阅者可以通过规范化格式定期访问其更新信息。

2.7.1 准备工作

本节实例的目标是读取一个 RSS 信息源并访问其中的一条内容。为此，我们将使用全球之声（Mashable）提供的 RSS 信息源。全球之声是一个数字媒体网站。简而言之，它是一个科技和社交媒体的博客列表。该网站的 RSS 信息源网址（URL）是 http://feeds.mashable.com/Mashable。另外，我们需要用 feedparser 库来读取 RSS 信息源。打开终端并运行以下命令即可在你的计算机上安装这个库：

```
pip install feedparser
```

安装好 feedparser 库后，我们就可以开始实现第一个读取 RSS 信息源的 Python 程序。

2.7.2 如何实现

1. 创建一个新的 Python 文件，命名为 rssReader.py，并添加以下代码：

```
import feedparser
```

2. 将全球之声信息源（Mashable feed）载入内存中，添加以下代码：

```
myFeed = feedparser.parse("http://feeds.mashable.com/Mashable")
```

myFeed 对象包含全球之声信息源的第一页，通过 feedparser 自动下载和解析该信息源并填充到合适的位置。myFeed 对象的条目列表将包含每个帖子（post）。

3. 检查当前信息源的标题并计算帖子数目：

```
print('Feed Title :', myFeed['feed']['title'])
print('Number of posts :', len(myFeed.entries))
```

在第一行代码中，我们通过 myFeed 对象获取到了信息源的标题。在第二行代码中，我们计算了 myFeed 对象中 entries 对象的长度。如前所述，entries 对象是一个包含解析后信息源中所有帖子的列表。运行代码，输出如下所示：

```
Feed Title: Mashable
Number of posts : 30
```

标题是 Mashable，当前，Mashable 每次最多存放 30 个帖子到信息源。

4. 从 entries 列表中获取第一个 post，并打印输出其标题：

```
post = myFeed.entries[0]
print('Post Title :',post.title)
```

在第一行代码中，我们获取了 entries 列表中的第一个元素并将其加载到 post 对象中。在第二行代码中，我们打印输出了 post 对象的标题。运行代码，输出应该与以下内容相似：

```
Post Title: The moon literally blocked the sun on Twitter
```

这里提到输出内容应该与其相似而不是完全一样，是因为信息源在不断自我更新。

5. 访问 post 的原始 HTML 内容，并将其打印输出：

```
content = post.content[0].value
print('Raw content :\n',content)
```

首先，我们访问 post 的内容对象并获取其具体值，打印输出如下：

```
Output: Feed Title: Mashable
Number of posts : 30
Post Title: The moon literally blocked the sun on Twitter
Raw content :
<img alt=""
src="https://i.amz.mshcdn.com/DzkxxIQCjyFHGoIBJoRGoYU3Y8o=/575x323/
filters:quality(90)/https%3A%2F%2Fblueprint-api-
production.s3.amazonaws.com%2Fuploads%2Fcard%2Fimage%2F569570%2F0ca
3e1bf-a4a2-4af4-85f0-1bbc8587014a.jpg" /><div style="float: right;
width: 50px;"><a
href="http://twitter.com/share?via=Mashable&text=The+moon+literally
+blocked+the+sun+on+Twitter&url=http%3A%2F%2Fmashable.com%2F2017%2F
08%2F21%2Fmoon-blocks-sun-eclipse-2017-
twitter%2F%3Futm_campaign%3DMash-Prod-RSS-Feedburner-All-
Partial%26utm_cid%3DMash-Prod-RSS-Feedburner-All-Partial"
style="margin: 10px;">
<p>The national space agency threw shade the best way it knows how:
by blocking the sun. Yep, you read that right. </p>
<div><div><blockquote>
<p>HA HA HA I've blocked the Sun! Make way for the Moon<a
href="https://twitter.com/hashtag/SolarEclipse2017?src=hash">#Solar
Eclipse2017</a> <a
href="https://t.co/nZCoqBlSTe">pic.twitter.com/nZCoqBlSTe</a></p>
<p>— NASA Moon (@NASAMoon) <a
href="https://twitter.com/NASAMoon/status/899681358737539073">Augus
t 21, 2017</a></p>
</blockquote></div></div>
```

2.7.3　工作原理

互联网上大多数的 RSS 信息源都以时间顺序排列，将最新的帖子放到最上面。因此，在该实例中我们每次访问的都是信息源提供的最新内容。信息源本身是不断更新的。所以，每次运行程序时，输出的格式保持不变，但是输出的内容却可能发生改变，这取决于信息源更新的速度。另外，我们在控制台直接输出原始的 HTML 文本而不是其文本内容。接下来，我们将解析 HTML 并从页面获取我们需要的信息。最后，本实例可以附加以下内容：读取你想要的任何信息源，将信息源中所有帖子的信息存储到磁盘，并利用它创建一个纯文本的语料库。当然，你可以从上一个和下一个实例中获得启发。

2.8　使用 BeautifulSoup 解析 HTML

大多数情况下，你需要处理的网上数据都以 HTML 页面的形式存在。因此，我们认为有必要向你介绍 Python 的 HTML 解析方法。有很多 Python 模块可以用来解析 HTML，在接下来的实例中，我们将使用 BeautifulSoup4 库来解析 HTML。

2.8.1　准备工作

BeautifulSoup4 包适用于 Python2 和 Python3。在使用这个包之前，我们需要提前下载并将它安装在解释器上。和之前一样，我们将使用 pip 来安装这个包。在命令行运行以下命令：

```
pip install beautifulsoup4
```

另外，你还需要本章 Dropbox 文件夹中的 sample-html.html 文件。如果你还没有下载该文件，请从以下链接下载：https://www.dropbox.com/sh/bk18dizhsu1p534/AABEuJw4TArUbzJf4Aa8gp5Wa?dl=0。

2.8.2　如何实现

1. 完成所有准备工作后，从导入以下声明开始：

```
from bs4 import BeautifulSoup
```

从 bs4 模块中导入 BeautifulSoup 类，它将用于解析 HTML。

2. 将一个 HTML 文件加载到 BeautifulSoup 对象中：

```
html_doc = open('sample-html.html', 'r').read()
soup = BeautifulSoup(html_doc, 'html.parser')
```

在第一行代码中，我们将 sample-html.html 文件的内容加载到 str 对象 html_doc 中。然后，创建了一个 BeautifulSoup 对象，需要解析的 HTML 文件作为第一个参数，html.parser 作为第二个参数。通过以上操作，BeautifulSoup 对象使用 html 解析器来解析文档。它将文档内容加载到 soup 对象中进行解析以备使用。

3. soup 对象最主要、最简单且最有用的功能就是去除所有的 HTML 标签并获取文本内容。添加以下代码：

```
print('\n\nFull text HTML Stripped:')
print(soup.get_text())
```

在 soup 对象上调用的 get_text() 方法将返回 HTML 标签去除后的文件内容。运行以上代码，将得到以下输出：

```
Full text HTML Stripped:
Sample Web Page

Main heading
This is a very simple HTML document
Improve your image by including an image.
Add a link to your favorite Web site.
This is a new sentence without a paragraph break, in bold italics.
This is purely the contents of our sample HTML document without any
of the HTML tags.
```

4. 有时不仅需要去除 HTML 标签，可能还需要获取特定标签的内容。访问其中的一个标签：

```
print('Accessing the <title> tag :', end=' ')
print(soup.title)
```

soup.title 将返回文件中的第一个标题（title）标签。以上代码的输出如下所示：

```
Accessing the <title> tag : <title>Sample Web Page</title>
```

5. 现在，我们需要某个 HTML 标签的文本内容。通过以下代码获取 <h1> 标签的内容：

```
print('Accessing the text of <H1> tag :', end=' ')
print(soup.h1.string)
```

soup.h1.string 命令将返回以 <h1> 标签开头的文本。以上代码的输出如下所示：

```
Accessing the text of <H1> tag : Main heading
```

6. 访问标签的属性。这里，我们将访问 img 标签的 alt 属性。添加以下代码行：

```
print('Accessing property of <img> tag :', end=' ')
print(soup.img['alt'])
```

通过仔细观察，你会发现访问标签属性的语法和访问标签文本的语法是不同的。运行以上代码，得到以下输出：

```
Accessing property of <img> tag : A Great HTML Resource
```

7. 最后，一个 HTML 文件中同一类型的标签可能多次出现。使用“.”语法仅能获取文件中第一次出现的标签。为了获取所有的标签，我们将使用 find_all() 函数，如下所示：

```
print('\nAccessing all occurences of the <p> tag :')
for p in soup.find_all('p'):
  print(p.string)
```

在 BeautifulSoup 对象上调用 find_all() 函数，参数是标签名，它将搜索整个 HTML 树并返回符合条件的标签列表。我们使用 for 循环来遍历该列表，并将 BeautifulSoup 对象中所有 <p> 标签的内容 / 文本打印并输出：

```
Output: Full text HTML Stripped:

Sample Web Page
```

```
Main heading
This is a very simple HTML document
Improve your image by including an image.

Add a link to your favorite Web site.
 This is a new sentence without a paragraph break, in bold italics.


Accessing the <title> tag : <title>Sample Web Page</title>
Accessing the text of <H1> tag : Main heading
Accessing property of <img> tag : A Great HTML Resource

Accessing all occurences of the <p> tag :
This is a very simple HTML document
Improve your image by including an image.
None
```

2.8.3 工作原理

BeautifulSoup4是一个很方便的库，可以用于解析任何HTML和XML内容。它支持Python内置的HTML解析器，但是你也可以使用其他第三方的解析器，例如，lxml解析器和纯Python的html5lib解析器。在本节中，我们使用Python内置的HTML解析器。如果你了解了HTML并会编写简单的HTML代码的话，输出结果是非常容易理解的。

CHAPTER 3

第3章 预处理

3.1 引言

在之前的章节，我们学习了如何读取、规范化和归一化来源于异构形式及格式的原始数据。在本章，我们将会进一步在 NLP 应用中准备和使用数据。在任何数据处理任务中，预处理都是最重要的步骤，否则无效的输入和输出将会严重影响计算机的处理效率。本章的目的是介绍一些关键的处理步骤，比如分词、词干提取和词形还原等。

在本章，我们将了解六部分不同的内容。按照预处理顺序，我们分别对分词、词干提取、词形还原、停用词处理和编辑距离以单个实例的形式进行详细的介绍。在最后一节，我们将举例说明如何运用所学的预处理技术去查找任意两个文本的共有词。

3.2 分词——学习使用 NLTK 内置的分词器

理解分词的含义，我们为什么要分词以及如何分词。

3.2.1 准备工作

首先我们来了解什么是分词。当一个文档或者一个长字符串需要处理的时候，你首先要做的是将它拆分成一个个单词和标点符号，我们称这个过程为分词。接下来我们将了解 NLTK 中可用分词器的类型以及它们的用法。

3.2.2 如何实现

1. 创建一个名为 tokenizer.py 的文件并添加如下代码：

```
from nltk.tokenize import LineTokenizer, SpaceTokenizer,
TweetTokenizer
from nltk import word_tokenize
```

本节导入四种不同类型的分词器以便测试。

2. 我们将从 LineTokernizer 开始介绍。添加以下两行代码:

```
lTokenizer = LineTokenizer();
print("Line tokenizer output :",lTokenizer.tokenize("My name is
Maximus Decimus Meridius, commander of the Armies of the North,
General of the Felix Legions and loyal servant to the true emperor,
Marcus Aurelius. \nFather to a murdered son, husband to a murdered
wife. \nAnd I will have my vengeance, in this life or the next."))
```

3. 顾名思义，该分词器应该将输入的字符串拆分成行（非句子）。让我们看看分词器的输出效果:

```
Line tokenizer output : ['My name is Maximus Decimus Meridius,
commander of the Armies of the North, General of the Felix Legions
and loyal servant to the true emperor, Marcus Aurelius. ', 'Father
to a murdered son, husband to a murdered wife. ', 'And I will have
my vengeance, in this life or the next.']
```

如上所示，它返回了一个包含三个字符串的列表。这意味着给定的输入已经根据换行符的位置被拆分成了三行。LineTokernizer 的作用是将输入的字符串拆分成行。

4. 现在我们来看 SpaceTokenizer。顾名思义，它是根据空格符来分词的。加入以下几行:

```
rawText = "By 11 o'clock on Sunday, the doctor shall open the
dispensary."
sTokenizer = SpaceTokenizer()
print("Space Tokenizer output :",sTokenizer.tokenize(rawText))
```

5. sTokenizer 是 SpaceTokenize 类的一个对象，调用 tokenize() 方法我们将看到如下输出:

```
Space Tokenizer output : ['By', '11', "o'clock", 'on', 'Sunday,',
'the', 'doctor', 'shall', 'open', 'the', 'dispensary.']
```

6. 正如期望的那样，输入的 rawText 被空格符 "" 拆分。接下来，调用 word_tokenize() 方法，示例如下:

```
print("Word Tokenizer output :", word_tokenize(rawText))
```

7. 我们可以看到这两者的区别。以上我们了解的 LineTokernizer 和 SpaceTokenizer 是 NLTK 模块中的类。接下来的大多数时间我们将使用这些类的方法，因为它实现的正是我们定义的分词任务。它可以对词语和标点符号进行拆分。运行结果如下:

```
Word Tokenizer output : ['By', '11', "o'clock", 'on', 'Sunday',
',', 'the', 'doctor', 'shall', 'open', 'the', 'dispensary', '.']
```

8. 如上所示，SpaceTokenizer 和 word_tokenize() 的区别是显而易见的。

9. 最后我们介绍一下 TweetTokernizer，处理特殊字符串的时候可以使用该分词器:

```
tTokenizer = TweetTokenizer()
print("Tweet Tokenizer output :",tTokenizer.tokenize("This is a
```

```
cooool #dummysmiley: :-) :-P <3"))
```

10. Tweets 包含我们想要保持完整的特殊单词、特殊字符、标签、笑脸符号等。上述代码的运行结果如下：

```
Tweet Tokenizer output : ['This', 'is', 'a', 'cooool',
'#dummysmiley', ':', ':-)', ':-P', '<3']
```

正如我们看到的，Tokenizer 保持特殊字符的完整性而没有进行拆分，笑脸符号也保持原封不动。这是一种特殊且比较少见的类，当需要的时候可以使用它。

11. 下面是程序运行的所有结果。我们已经进行了详细的介绍，就不再赘述了。

```
Line tokenizer output : ['My name is Maximus Decimus Meridius,
commander of the Armies of the North, General of the Felix Legions
and loyal servant to the true emperor, Marcus Aurelius. ', 'Father
to a murdered son, husband to a murdered wife. ', 'And I will have
my vengeance, in this life or the next.']
Space Tokenizer output : ['By', '11', "o'clock", 'on', 'Sunday,',
'the', 'doctor', 'shall', 'open', 'the', 'dispensary.']
Word Tokenizer output : ['By', '11', "o'clock", 'on', 'Sunday',
',', 'the', 'doctor', 'shall', 'open', 'the', 'dispensary', '.']
Tweet Tokenizer output : ['This', 'is', 'a', 'cooool',
'#dummysmiley', ':', ':-)', ':-P', '<3']
```

3.2.3 工作原理

我们学习了 NLTK 库中实现分词的三个分词器类和一个方法。理解它们的工作原理并不是一件特别困难的事，但理解它的运行原理是非常有价值的。语言处理任务的最小处理单元是词。这非常像一种分而治之的策略，我们首先理解粒度级别上最小的单元的语义，然后通过语义叠加来理解句子、段落、文档和语料库（如果有的话）的语义。

3.3 词干提取——学习使用 NLTK 内置的词干提取器

在本节我们需要了解词干的概念及词干提取的过程。我们将了解词干提取的意义以及如何使用 NLTK 内置的词干提取器。

3.3.1 准备工作

首先，词干是什么？词干是没有任何后缀的词的基本组成部分，词干提取器的作用是去除后缀并输出词的词干。下面我们介绍 NLTK 中几种类型的词干提取器。

3.3.2 如何实现

1. 创建一个名为 stemmers.py 的文件，并添加以下导入行：

```
from nltk import PorterStemmer, LancasterStemmer, word_tokenize
```

在本节中我们导入两种不同类型的词干提取。

2. 在进行词干提取之前，我们首先需要对输入文本进行分词，使用以下代码来完成这一步：

```
raw = "My name is Maximus Decimus Meridius, commander of the Armies
of the North, General of the Felix Legions and loyal servant to the
true emperor, Marcus Aurelius. Father to a murdered son, husband to
a murdered wife. And I will have my vengeance, in this life or the
next."
tokens = word_tokenize(raw)
```

分词列表包含输入字符串raw产生的所有分词。

3. 首先使用PorterStemmer，添加如下三行代码：

```
porter = PorterStemmer()
pStems = [porter.stem(t) for t in tokens]
print(pStems)
```

4. 首先初始化词干提取器，然后对所有的输入文本应用该词干提取器，最后打印输出结果。通过观察输出结果，我们可以了解到更多信息：

```
['My', 'name', 'is', 'maximu', 'decimu', 'meridiu', ',', 'command',
'of', 'the', 'armi', 'of', 'the', 'north', ',', 'gener', 'of',
'the', 'felix', 'legion', 'and', 'loyal', 'servant', 'to', 'the',
'true', 'emperor', ',', 'marcu', 'aureliu', '.', 'father', 'to',
'a', 'murder', 'son', ',', 'husband', 'to', 'a', 'murder', 'wife',
'.', 'and', 'I', 'will', 'have', 'my', 'vengeanc', ',', 'in',
'thi', 'life', 'or', 'the', 'next', '.']
```

正如你在输出结果中看到的，所有的单词都去除了“s”“es”“e”“ed”“al”等后缀。

5. 接下来使用LancasterStemmer，与porter相比较，它更容易出错，因为它包含更多要去除的尾缀：

```
lancaster = LancasterStemmer()
lStems = [lancaster.stem(t) for t in tokens]
print(lStems)
```

6. 进行相似实验，用LancasterStemmer代替PorterStemmer。输出结果如下：

```
['my', 'nam', 'is', 'maxim', 'decim', 'meridi', ',', 'command',
'of', 'the', 'army', 'of', 'the', 'nor', ',', 'gen', 'of', 'the',
'felix', 'leg', 'and', 'loy', 'serv', 'to', 'the', 'tru', 'emp',
',', 'marc', 'aureli', '.', 'fath', 'to', 'a', 'murd', 'son', ',',
'husband', 'to', 'a', 'murd', 'wif', '.', 'and', 'i', 'wil', 'hav',
'my', 'veng', ',', 'in', 'thi', 'lif', 'or', 'the', 'next', '.']
```

我们将在输出部分讨论它们的差别，但是很容易就能看出该分词器对尾缀的处理优于Porter。尾缀如“us”“e”“th”“eral”“ered”等。

7. 下面是程序所有的输出。比较两种词干提取器的输出：

```
['My', 'name', 'is', 'maximu', 'decimu', 'meridiu', ',', 'command',
'of', 'the', 'armi', 'of', 'the', 'north', ',', 'gener', 'of',
'the', 'felix', 'legion', 'and', 'loyal', 'servant', 'to', 'the',
```

```
'true', 'emperor', ',', 'marcu', 'aureliu', '.', 'father', 'to',
'a', 'murder', 'son', ',', 'husband', 'to', 'a', 'murder', 'wife',
'.', 'and', 'I', 'will', 'have', 'my', 'vengeanc', ',', 'in',
'thi', 'life', 'or', 'the', 'next', '.']
['my', 'nam', 'is', 'maxim', 'decim', 'meridi', ',', 'command',
'of', 'the', 'army', 'of', 'the', 'nor', ',', 'gen', 'of', 'the',
'felix', 'leg', 'and', 'loy', 'serv', 'to', 'the', 'tru', 'emp',
',', 'marc', 'aureli', '.', 'fath', 'to', 'a', 'murd', 'son', ',',
'husband', 'to', 'a', 'murd', 'wif', '.', 'and', 'i', 'wil', 'hav',
'my', 'veng', ',', 'in', 'thi', 'lif', 'or', 'the', 'next', '.']
```

通过比较这两种词干提取器的输出，我们发现在去除尾缀方面 lancaster 做得更加彻底。它尽可能多地去除尾部字符，而 porter 相对来说尽可能少地去除尾部字符。

3.3.3 工作原理

对于一些语言处理任务，我们需要忽略输入文本的格式，而只需对其词干进行处理。例如，当你在因特网上搜索 cameras 时，搜索结果将包括与 camera 以及 cameras 相关的文档，反之亦然。通过这个例子我们知道，这两个词的词干其实是一样的，都是 camera。

我们可以明显地知道词干提取很容易造成问题，当词干提取器提取单词词干后，单词拼写被改变了。即使偶尔几次，它可能不会造成问题，但当我们想理解语义的时候，词干提取会丢失许多数据。基于这个原因，接下来我们需要学习什么是**词形还原**。

3.4 词形还原——学习使用 NLTK 中的 WordnetLemmatizer 函数

本节学习什么是词元（lemma）及词形还原（lemmatization）。了解词形还原与词干提取的差异、词形还原的重要性，以及如何使用 NLTK 库中的 WordnetLemmatizer 函数来完成任务。

3.4.1 准备工作

一个词元是一个词的中心词，或者简单地说是一个词的基本组成。我们已经了解了什么是词干，但是与词干提取过程不同的是，词干是通过去除或替换尾缀获得的，而词元获取是一个字典匹配过程。由于词形还原是一个字典映射过程，因此词形还原相对于词干提取来说，是一个更复杂的过程。

3.4.2 如何工作

1. 创建一个名为 lemmatizer.py 的文件并添加如下代码：

```
from nltk import word_tokenize, PorterStemmer, WordNetLemmatizer
```

首先需要对这个句子进行分词，这里我们将使用 PorterStemmer 来比较输出。

2. 在进行任何词干提取之前，我们首先需要对输入文本进行分词，使用如下代码来完成：

```
raw = "My name is Maximus Decimus Meridius, commander of the armies
of the north, General of the Felix legions and loyal servant to the
true emperor, Marcus Aurelius. Father to a murdered son, husband to
a murdered wife. And I will have my vengeance, in this life or the
next."
tokens = word_tokenize(raw)
```

分词列表包含输入字符串 raw 产生的所有单词。

3. 首先我们使用 PorterStemmer，在上一节我们已经见过这个函数。添加如下三行：

```
porter = PorterStemmer()
stems = [porter.stem(t) for t in tokens]
print(stems)
```

第一步初始化词干提取器，然后对输入文本应用该词干提取器，最后打印输出。在本节的最后，我们将检验输出。

4. 现在我们使用词形还原器 lemmatizer，添加如下三行代码：

```
lemmatizer = WordNetLemmatizer()
lemmas = [lemmatizer.lemmatize(t) for t in tokens]
print(lemmas)
```

5. 运行程序，上述三行代码的输出如下所示：

```
['My', 'name', 'is', 'Maximus', 'Decimus', 'Meridius', ',',
'commander', 'of', 'the', 'army', 'of', 'the', 'north', ',',
'General', 'of', 'the', 'Felix', 'legion', 'and', 'loyal',
'servant', 'to', 'the', 'true', 'emperor', ',', 'Marcus',
'Aurelius', '.', 'Father', 'to', 'a', 'murdered', 'son', ',',
'husband', 'to', 'a', 'murdered', 'wife', '.', 'And', 'I', 'will',
'have', 'my', 'vengeance', ',', 'in', 'this', 'life', 'or', 'the',
'next', '.']
```

如上所示，词形还原器能够判断专有名词，其不需要去除尾缀“s”，但对于非专有名词（如 legions 和 armies）来说需要去除并替换其尾缀。词形还原是一个字典匹配过程，我们将在输出部分讨论词干提取与词形还原的区别。

6. 下面是程序的全部输出结果。我们比较词干提取器与词形还原器的输出：

```
['My', 'name', 'is', 'maximu', 'decimu', 'meridiu', ',', 'command',
'of', 'the', 'armi', 'of', 'the', 'north', ',', 'gener', 'of',
'the', 'felix', 'legion', 'and', 'loyal', 'servant', 'to', 'the',
'true', 'emperor', ',', 'marcu', 'aureliu', '.', 'father', 'to',
'a', 'murder', 'son', ',', 'husband', 'to', 'a', 'murder', 'wife',
'.', 'and', 'I', 'will', 'have', 'my', 'vengeanc', ',', 'in',
'thi', 'life', 'or', 'the', 'next', '.']
['My', 'name', 'is', 'Maximus', 'Decimus', 'Meridius', ',',
'commander', 'of', 'the', 'army', 'of', 'the', 'north', ',',
'General', 'of', 'the', 'Felix', 'legion', 'and', 'loyal',
'servant', 'to', 'the', 'true', 'emperor', ',', 'Marcus',
'Aurelius', '.', 'Father', 'to', 'a', 'murdered', 'son', ',',
'husband', 'to', 'a', 'murdered', 'wife', '.', 'And', 'I', 'will',
'have', 'my', 'vengeance', ',', 'in', 'this', 'life', 'or', 'the',
'next', '.']
```

通过比较词干提取器和词形还原器的输出，我们发现词干提取器造成许多错误，而词形还原器只造成很少的错误，但是它没有对单词 murdered 做任何处理，这是一个处理错误。从最后的结果来看，在提取词的基础形式方面，词形还原器比词干提取器表现更优。

3.4.3 工作原理

仅当 WordnetLemmatizer 能在字典中查找到目标词时，它才会去除词缀。这使得词形还原的处理速度比词干提取的处理速度慢。而且，它识别首字母大写的单词并将其作为特殊词。它对这些特殊词不做任何处理，只是将它们按原样返回。为了避免这个问题，你可能需要将你的输入字符串转换成小写字母，然后再执行词形还原。

虽然如此，词形还原仍然不是完美的，它也存在错误。检查本实例的输入与输出结果，我们发现它不能将 murdered 转换成 murder。类似地，它能够正确处理 women 这个词，但却不能正确处理 men 这个词。

3.5 停用词——学习使用停用词语料库及其应用

在本节中，我们将以古登堡语料库（Gutenburg Corpus）为例子。古登堡语料库是 NLTK 数据模块中的一部分。它从古登堡文件档案中大约 25 000 本电子书中选取了 18 个文本。它是纯文本语料库（PlainTextCorpus），因为该语料库没有进行任何分类，所以非常适用于单纯的词处理，而不需考虑其与任何主题的相关性。本节的目标之一就是介绍文本分析过程中最重要的预处理过程——停用词处理。

根据目标，我们将通过这个语料库阐述 Python 的 NLTK 模块中频率分布和停用词语料库的应用。简而言之，停用词是一种具有很少语义价值，却具有极高句法价值的单词。当你不进行句法分析，而使用词袋（bag-of-word）方法的时候（例如 TF/IDF），通常需要去除停用词。

3.5.1 准备工作

本节我们将使用 nltk.corpus.stopwords 和 nltk.corpus.gutenberg，它们是 NLTK 数据模块的一部分。

3.5.2 如何工作

1. 创建一个名为 Gutenberg.py 的文件并导入如下三行代码：

```
import nltk
from nltk.corpus import gutenberg
print(gutenberg.fileids())
```

2. 前两行代码导入 Gutenberg 语料库及其他需要的语料库，第三行代码用于检查是否成功加载语料库。在 Python 集成环境中运行这个文件，输出的结果如下：

```
['austen-emma.txt', 'austen-persuasion.txt', 'austen-sense.txt',
'bible-kjv.txt', 'blake-poems.txt', 'bryant-stories.txt', 'burgess-
busterbrown.txt', 'carroll-alice.txt', 'chesterton-ball.txt',
'chesterton-brown.txt', 'chesterton-thursday.txt', 'edgeworth-
parents.txt', 'melville-moby_dick.txt', 'milton-paradise.txt',
'shakespeare-caesar.txt', 'shakespeare-hamlet.txt', 'shakespeare-
macbeth.txt', 'whitman-leaves.txt']
```

如上所示，18 个古登堡文件的名称都被打印到了屏幕上。

3. 添加如下两行代码，我们对语料库中的所有单词列表做一些简单的预处理：

```
gb_words = gutenberg.words('bible-kjv.txt')
words_filtered = [e for e in gb_words if len(e) >= 3]
```

第一行代码拷贝了语料库样例 bible-kjv.txt 的所有单词列表，并存储在 gb-words 变量中。第二行代码遍历古登堡语料库的所有单词列表，去除所有长度小于 3 的单词。

4. 现在，我们使用 nltk.corpus.stopwords，并在之前过滤后的单词列表上做停用词处理。添加如下几行代码：

```
stopwords = nltk.corpus.stopwords.words('english')
words = [w for w in words_filtered if w.lower() not in stopwords]
```

第一行代码简单地从停用词语料库中加载了英文（english）停用词到 stopwords 变量中。第二行代码是我们对之前过滤后的单词词表做进一步处理，过滤掉所有停用词。

5. 现在我们分别在预处理后的词表 words 和未做处理的词表 gb_words 上分别应用 nltk.FreqDist，加入如下几行代码：

```
fdistPlain = nltk.FreqDist(words)
fdist = nltk.FreqDist(gb_words)
```

通过传递我们在步骤 3 和 4 中得到的单词列表作为参数，创建 FreqDist 对象。

6. 如果我们想看到执行上述操作后的频率分布特征，加入下面四行代码，我们将会看到如下结果：

```
print('Following are the most common 10 words in the bag')
print(fdistPlain.most_common(10))
print('Following are the most common 10 words in the bag minus the
stopwords')
print(fdist.most_common(10))
```

most_common(10) 函数将返回被词频分布处理后的词袋中的 10 个最常用的词，之后我们将讨论和阐述这些输出结果。

7. 运行上述程序后，你将得到类似如下的输出：

```
['austen-emma.txt', 'austen-persuasion.txt', 'austen-sense.txt',
'bible-kjv.txt', 'blake-poems.txt', 'bryant-stories.txt', 'burgess-
busterbrown.txt', 'carroll-alice.txt', 'chesterton-ball.txt',
'chesterton-brown.txt', 'chesterton-thursday.txt', 'edgeworth-
parents.txt', 'melville-moby_dick.txt', 'milton-paradise.txt',
```

```
'shakespeare-caesar.txt', 'shakespeare-hamlet.txt', 'shakespeare-macbeth.txt', 'whitman-leaves.txt']

Following are the most common 10 words in the bag

[(',', 70509), ('the', 62103), (':', 43766), ('and', 38847), ('of', 34480), ('.', 26160), ('to', 13396), ('And', 12846), ('that', 12576), ('in', 12331)]

Following are the most common 10 words in the bag minus the stopwords

[('shall', 9838), ('unto', 8997), ('lord', 7964), ('thou', 5474),
('thy', 4600), ('god', 4472), ('said', 3999), ('thee', 3827),
('upon', 2748), ('man', 2735)]
```

3.5.3　工作原理

如果仔细观察结果，你将会发现：未处理文本中的10个最常见词并没有什么意义。而另一方面，预处理后的文本中的10个最常见词，例如god、lord以及man，向我们提示——我们正在处理与信仰或宗教有关的文本。

本节最重要的目标是向你介绍停用词的处理，这是在进行任何复杂文本数据分析之前都需要掌握的预处理技术。NLTK的stopwords语料库包含了11种语言，在任何文本分析应用中，当你需要分析关键词时，正确处理停用词将让你事半功倍。词频分布将帮助你获取重要的词语。从统计学角度来看，如果你绘制一个包含词频和词语重要度的二维平面图，理想的分布曲线看起来会像一条钟形曲线。

3.6　编辑距离——编写计算两个字符串之间编辑距离的算法

编辑距离（Edit distance），也被称为**Levenshtein距离**，是一种衡量两个字符串之间相似度的度量方法。实质上，它是一系列编辑操作的计数，包括删除、插入或替换操作，这些操作能够将字符串A转换为字符串B。我们编写算法来计算编辑距离，并将其和nltk.metrics.distance.edit_distance()进行准确性比较和检验。

3.6.1　准备工作

你可能想了解Levenshtein距离相关的数学公式，但这里我们只讨论Python的算法实现过程及算法原理。这些并不能完全让人理解它背后的数学原理，因此这里提供一个可供参考的维基百科地址：https://en.wikipedia.Org/wiki/Levenshtein_distance。

3.6.2　如何实现

1. 创建一个名为edit_distance_calculator.py的文件，并输入下面的代码：

```
from nltk.metrics.distance import edit_distance
```

我们导入了 NLTK 内置的包，包括 nltk.metrics.distance 模块中的 edit_distance 函数。

2. 定义一个函数接收两个字符串 str1 和 str2 为输入，并计算这两个字符串的编辑距离，然后返回一个整型的距离值：

```
def my_edit_distance(str1, str2):
```

3. 接下来获取两个输入字符串的长度，并使用这个长度创建一个 m x n 表，其中 m 和 n 分别表示字符串 str1 和 str2 的长度。

```
m=len(str1)+1
n=len(str2)+1
```

4. 创建一个 table 并初始化第一行和第一列：

```
table = {}
for i in range(m): table[i,0]=i
for j in range(n): table[0,j]=j
```

5. 初始化这个二维数组，其在内存中的存储内容如下表所示：

	0	1	2	3	4	
0	0	1	2	3	4	
1	1					A
2	2					N
3	3					D
		H	A	N	D	

请注意：这些内容都包含在函数内部，我们以这两个字符串作为范例是为了更好地解释算法。

6. 现在我们使用下面的公式来填充这些矩阵：

```
for i in range(1, m):
  for j in range(1, n):
    cost = 0 if str1[i-1] == str2[j-1] else 1
    table[i,j] = min(table[i, j-1]+1, table[i-1, j]+1, table[i-1,
j-1]+cost)
```

当两个字符串相同或需要编辑操作时，比如删除或插入，可以通过计算得到 cost 值。下一行公式用于计算矩阵中单元格的值，其中前两个表示替换操作的次数，第三个除了替换操作次数，我们还加上了之前步骤计算获得的 cost 值，然后从它们三者中取最小值。

7. 最后，返回最后一个单元格的值 table[m,n] 作为最终的编辑距离：

```
return table[i,j]
```

8. 现在调用我们的函数以及 NLTK 包中的 edit_distance() 函数来分别计算两个字符串的编辑距离，并输出结果：

```
print("Our Algorithm :",my_edit_distance("hand", "and"))
print("NLTK Algorithm :",edit_distance("hand", "and"))
```

9. 我们的单词是 hand 和 and。在第一个字符串上一个简单的删除操作或者在第二个字符串上一个简单的插入操作都能使它们匹配。因此，Levenshtein 的期望值是 1。

10. 这个程序的输出如下：

```
Our Algorithm : 1
NLTK Algorithm : 1
```

正如所期望的那样，NLTK 中 edit_distance() 函数以及我们定义的函数都返回 1，客观来说，我们的算法能够达到预期结果，但我仍然希望你们通过更多的例子进行测试。

3.6.3　工作原理

我们已经介绍了一小段简单的算法，现在我们来看看这个矩阵表是如何根据算法进行变化的。附表如下：

	0	1	2	3	4	
0	0	1	2	3	4	
1	1	1	1	2	3	A
2	2	2	2	1	2	N
3	3	3	3	2	1	D
		H	A	N	D	

你已经知道了如何初始化矩阵，现在我们使用算法中的公式来填充矩阵。浅灰色部分是重要的数字，在第一次迭代后，你可以看到编辑距离都是朝着 1 的方向进行移动，最终返回的是深灰色部分的值。

现在，编辑距离算法的应用是多种多样的。最常见的是，它被应用于拼写检查和对文本编辑者的自动提示，以及搜索引擎等众多诸如此类的文本处理任务。由于这种字符串比较的计算代价等同于要比较的两个字符串的长度乘积，所以有时用它来计算大文本的相似度是不切实际的。

3.7　处理两篇短文并提取共有词汇

当你遇到一个典型的文本分析问题时，本节会提供一个解决问题的方法，我们将使用多种预处理技术来得到输出。本节将介绍一个重要的预处理任务，但它并不是一个真实的文本分析应用。我们将使用从 http://www.english-for-students.com/ 下载的一些短文。

3.7.1　准备工作

本节我们将去除所有的特殊字符，进行分词，处理大小写以及对一些集合和列表进行操作。我们不会使用任何特殊的库，仅涉及 Python 自带的库。

3.7.2 如何操作

1. 创建一个名为 lemmatizer.py 的文件并创建一组包含长字符串的短文或新闻文章：

```
story1 = """In a far away kingdom, there was a river. This river
was home to many golden swans. The swans spent most of their time
on the banks of the river. Every six months, the swans would leave
a golden feather as a fee for using the lake. The soldiers of the
kingdom would collect the feathers and deposit them in the royal
treasury.
One day, a homeless bird saw the river. "The water in this river
seems so cool and soothing. I will make my home here," thought the
bird.
As soon as the bird settled down near the river, the golden swans
noticed her. They came shouting. "This river belongs to us. We pay
a golden feather to the King to use this river. You can not live
here."
"I am homeless, brothers. I too will pay the rent. Please give me
shelter," the bird pleaded. "How will you pay the rent? You do not
have golden feathers," said the swans laughing. They further added,
"Stop dreaming and leave once." The humble bird pleaded many times.
But the arrogant swans drove the bird away.
"I will teach them a lesson!" decided the humiliated bird.
She went to the King and said, "O King! The swans in your river are
impolite and unkind. I begged for shelter but they said that they
had purchased the river with golden feathers."
The King was angry with the arrogant swans for having insulted the
homeless bird. He ordered his soldiers to bring the arrogant swans
to his court. In no time, all the golden swans were brought to the
King's court.
"Do you think the royal treasury depends upon your golden feathers?
You can not decide who lives by the river. Leave the river at once
or you all will be beheaded!" shouted the King.
The swans shivered with fear on hearing the King. They flew away
never to return. The bird built her home near the river and lived
there happily forever. The bird gave shelter to all other birds in
the river. """
story2 = """Long time ago, there lived a King. He was lazy and
liked all the comforts of life. He never carried out his duties as
a King. "Our King does not take care of our needs. He also ignores
the affairs of his kingdom." The people complained.
One day, the King went into the forest to hunt. After having
wandered for quite sometime, he became thirsty. To his relief, he
spotted a lake. As he was drinking water, he suddenly saw a golden
swan come out of the lake and perch on a stone. "Oh! A golden swan.
I must capture it," thought the King.
But as soon as he held his bow up, the swan disappeared. And the
King heard a voice, "I am the Golden Swan. If you want to capture
me, you must come to heaven."
Surprised, the King said, "Please show me the way to heaven." "Do
good deeds, serve your people and the messenger from heaven would
come to fetch you to heaven," replied the voice.
The selfish King, eager to capture the Swan, tried doing some good
deeds in his Kingdom. "Now, I suppose a messenger will come to take
me to heaven," he thought. But, no messenger came.
The King then disguised himself and went out into the street. There
he tried helping an old man. But the old man became angry and said,
```

```
"You need not try to help. I am in this miserable state because of
out selfish King. He has done nothing for his people."
Suddenly, the King heard the golden swan's voice, "Do good deeds and
you will come to heaven." It dawned on the King that by doing
selfish acts, he will not go to heaven.
He realized that his people needed him and carrying out his duties
was the only way to heaven. After that day he became a responsible
King.
"""
```

这是从上面提及的网址上获取的两篇短文。

2. 第一步，删除文中的一些特殊字符。我们去除了所有换行符“\n”、逗号、句号、感叹号、问号等。最后，用 casefold() 函数将所有的字符串转换为小写：

```
story1 = story1.replace(",", "").replace("\n", "").replace('.',
'').replace('"', '').replace("!","").replace("?","").casefold()
story2 = story2.replace(",", "").replace("\n", "").replace('.',
'').replace('"', '').replace("!","").replace("?","").casefold()
```

3. 下一步，对文本进行分词：

```
story1_words = story1.split(" ")
print("First Story words :",story1_words)
story2_words = story2.split(" ")
print("Second Story words :",story2_words)
```

4. 对 story1 和 story2 调用 split，根据 "" 字符进行分词，得到它们的单词列表。现在来看该步骤的输出：

```
First Story words : ['in', 'a', 'far', 'away', 'kingdom', 'there',
'was', 'a', 'river', 'this', 'river', 'was', 'home', 'to', 'many',
'golden', 'swans', 'the', 'swans', 'spent', 'most', 'of', 'their',
'time', 'on', 'the', 'banks', 'of', 'the', 'river', 'every', 'six',
'months', 'the', 'swans', 'would', 'leave', 'a', 'golden',
'feather', 'as', 'a', 'fee', 'for', 'using', 'the', 'lake', 'the',
'soldiers', 'of', 'the', 'kingdom', 'would', 'collect', 'the',
'feathers', 'and', 'deposit', 'them', 'in', 'the', 'royal',
'treasury', 'one', 'day', 'a', 'homeless', 'bird', 'saw', 'the',
'river', 'the', 'water', 'in', 'this', 'river', 'seems', 'so',
'cool', 'and', 'soothing', 'i', 'will', 'make', 'my', 'home',
'here', 'thought', 'the', 'bird', 'as', 'soon', 'as', 'the',
'bird', 'settled', 'down', 'near', 'the', 'river', 'the', 'golden',
'swans', 'noticed', 'her', 'they', 'came', 'shouting', 'this',
'river', 'belongs', 'to', 'us', 'we', 'pay', 'a', 'golden',
'feather', 'to', 'the', 'king', 'to', 'use', 'this', 'river',
'you', 'can', 'not', 'live', 'here', 'i', 'am', 'homeless',
'brothers', 'i', 'too', 'will', 'pay', 'the', 'rent', 'please',
'give', 'me', 'shelter', 'the', 'bird', 'pleaded', 'how', 'will',
'you', 'pay', 'the', 'rent', 'you', 'do', 'not', 'have', 'golden',
'feathers', 'said', 'the', 'swans', 'laughing', 'they', 'further',
'added', 'stop', 'dreaming', 'and', 'leave', 'once', 'the',
'humble', 'bird', 'pleaded', 'many', 'times', 'but', 'the',
'arrogant', 'swans', 'drove', 'the', 'bird', 'away', 'i', 'will',
'teach', 'them', 'a', 'lesson', 'decided', 'the', 'humiliated',
'bird', 'she', 'went', 'to', 'the', 'king', 'and', 'said', 'o',
'king', 'the', 'swans', 'in', 'your', 'river', 'are', 'impolite',
```

```
'and', 'unkind', 'i', 'begged', 'for', 'shelter', 'but', 'they',
'said', 'that', 'they', 'had', 'purchased', 'the', 'river', 'with',
'golden', 'feathers', 'the', 'king', 'was', 'angry', 'with', 'the',
'arrogant', 'swans', 'for', 'having', 'insulted', 'the',
'homeless', 'bird', 'he', 'ordered', 'his', 'soldiers', 'to',
'bring', 'the', 'arrogant', 'swans', 'to', 'his', 'court', 'in',
'no', 'time', 'all', 'the', 'golden', 'swans', 'were', 'brought',
'to', 'the', 'king's', 'court', 'do', 'you', 'think', 'the',
'royal', 'treasury', 'depends', 'upon', 'your', 'golden',
'feathers', 'you', 'can', 'not', 'decide', 'who', 'lives', 'by',
'the', 'river', 'leave', 'the', 'river', 'at', 'once', 'or', 'you',
'all', 'will', 'be', 'beheaded', 'shouted', 'the', 'king', 'the',
'swans', 'shivered', 'with', 'fear', 'on', 'hearing', 'the',
'king', 'they', 'flew', 'away', 'never', 'to', 'return', 'the',
'bird', 'built', 'her', 'home', 'near', 'the', 'river', 'and',
'lived', 'there', 'happily', 'forever', 'the', 'bird', 'gave',
'shelter', 'to', 'all', 'other', 'birds', 'in', 'the', 'river', '']
Second Story words : ['long', 'time', 'ago', 'there', 'lived', 'a',
'king', 'he', 'was', 'lazy', 'and', 'liked', 'all', 'the',
'comforts', 'of', 'life', 'he', 'never', 'carried', 'out', 'his',
'duties', 'as', 'a', 'king', '“our', 'king', 'does', 'not', 'take',
'care', 'of', 'our', 'needs', 'he', 'also', 'ignores', 'the',
'affairs', 'of', 'his', 'kingdom', 'the', 'people', 'complained',
'one', 'day', 'the', 'king', 'went', 'into', 'the', 'forest', 'to',
'hunt', 'after', 'having', 'wandered', 'for', 'quite', 'sometime',
'he', 'became', 'thirsty', 'to', 'his', 'relief', 'he', 'spotted',
'a', 'lake', 'as', 'he', 'was', 'drinking', 'water', 'he',
'suddenly', 'saw', 'a', 'golden', 'swan', 'come', 'out', 'of',
'the', 'lake', 'and', 'perch', 'on', 'a', 'stone', '“oh', 'a',
'golden', 'swan', 'i', 'must', 'capture', 'it', 'thought', 'the',
'king', 'but', 'as', 'soon', 'as', 'he', 'held', 'his', 'bow',
'up', 'the', 'swan', 'disappeared', 'and', 'the', 'king', 'heard',
'a', 'voice', '“i', 'am', 'the', 'golden', 'swan', 'if', 'you',
'want', 'to', 'capture', 'me', 'you', 'must', 'come', 'to',
'heaven', 'surprised', 'the', 'king', 'said', '“please', 'show',
'me', 'the', 'way', 'to', 'heaven', '“do', 'good', 'deeds',
'serve', 'your', 'people', 'and', 'the', 'messenger', 'from',
'heaven', 'would', 'come', 'to', 'fetch', 'you', 'to', 'heaven',
'replied', 'the', 'voice', 'the', 'selfish', 'king', 'eager', 'to',
'capture', 'the', 'swan', 'tried', 'doing', 'some', 'good',
'deeds', 'in', 'his', 'kingdom', '“now', 'i', 'suppose', 'a',
'messenger', 'will', 'come', 'to', 'take', 'me', 'to', 'heaven',
'he', 'thought', 'but', 'no', 'messenger', 'came', 'the', 'king',
'then', 'disguised', 'himself', 'and', 'went', 'out', 'into',
'the', 'street', 'there', 'he', 'tried', 'helping', 'an', 'old',
'man', 'but', 'the', 'old', 'man', 'became', 'angry', 'and',
'said', '“you', 'need', 'not', 'try', 'to', 'help', 'i', 'am',
'in', 'this', 'miserable', 'state', 'because', 'of', 'out',
'selfish', 'king', 'he', 'has', 'done', 'nothing', 'for', 'his',
'people', 'suddenly', 'the', 'king', 'heard', 'the', 'golden',
'swan's', 'voice', '“do', 'good', 'deeds', 'and', 'you', 'will',
'come', 'to', 'heaven', 'it', 'dawned', 'on', 'the', 'king',
'that', 'by', 'doing', 'selfish', 'acts', 'he', 'will', 'not',
'go', 'to', 'heaven', 'he', 'realized', 'that', 'his', 'people',
'needed', 'him', 'and', 'carrying', 'out', 'his', 'duties', 'was',
'the', 'only', 'way', 'to', 'heaven', 'after', 'that', 'day', 'he',
'became', 'a', 'responsible', 'king', '']
```

正如你所看到的，所有的特殊字符都被去除，并创建了一个单词列表。

5. 现在，我们基于这个单词列表创建一个词汇集。一个词汇集是由一组非重复的单词组成的：

```
story1_vocab = set(story1_words)
print("First Story vocabulary :",story1_vocab)
story2_vocab = set(story2_words)
```

6. 调用 Python 自带的 set() 函数将这个列表转换成一个集合：

```
First Story vocabulary : {'', 'king's', 'am', 'further', 'having',
'river', 'he', 'all', 'feathers', 'banks', 'at', 'shivered',
'other', 'are', 'came', 'here', 'that', 'soon', 'lives', 'unkind',
'by', 'on', 'too', 'kingdom', 'never', 'o', 'make', 'every',
'will', 'said', 'birds', 'teach', 'away', 'hearing', 'humble',
'but', 'deposit', 'them', 'would', 'leave', 'return', 'added',
'were', 'fear', 'bird', 'lake', 'my', 'settled', 'or', 'pleaded',
'in', 'so', 'use', 'was', 'me', 'us', 'laughing', 'bring', 'rent',
'have', 'how', 'lived', 'of', 'seems', 'gave', 'day', 'no',
'months', 'down', 'this', 'the', 'her', 'decided', 'angry',
'built', 'cool', 'think', 'golden', 'spent', 'time', 'noticed',
'lesson', 'many', 'near', 'once', 'collect', 'who', 'your', 'flew',
'fee', 'six', 'most', 'had', 'to', 'please', 'purchased',
'happily', 'depends', 'belongs', 'give', 'begged', 'there', 'she',
'i', 'times', 'dreaming', 'as', 'court', 'their', 'you', 'shouted',
'shelter', 'forever', 'royal', 'insulted', 'they', 'with', 'live',
'far', 'water', 'king', 'shouting', 'a', 'brothers', 'drove',
'arrogant', 'saw', 'soldiers', 'stop', 'home', 'upon', 'can',
'decide', 'beheaded', 'do', 'for', 'homeless', 'ordered', 'be',
'using', 'not', 'feather', 'soothing', 'swans', 'humiliated',
'treasury', 'thought', 'one', 'and', 'we', 'impolite', 'brought',
'went', 'pay', 'his'}
Second Story vocabulary {'', 'needed', 'having', 'am', 'he', 'all',
'way', 'spotted', 'voice', 'realized', 'also', 'came', 'that',
'"our', 'soon', '"oh', 'by', 'on', 'has', 'complained', 'never',
'ago', 'kingdom', '"do', 'capture', 'said', 'into', 'long', 'will',
'liked', 'disappeared', 'but', 'would', 'must', 'stone', 'lake',
'from', 'messenger', 'eager', 'deeds', 'fetch', 'carrying', 'in',
'because', 'perch', 'responsible', 'was', 'me', 'disguised',
'take', 'comforts', 'lived', 'of', 'tried', 'day', 'no', 'street',
'good', 'bow', 'the', 'need', 'this', 'helping', 'angry', 'out',
'thirsty', 'relief', 'wandered', 'old', 'golden', 'acts', 'time',
'an', 'needs', 'suddenly', 'state', 'serve', 'affairs', 'ignores',
'does', 'people', 'want', 'your', 'dawned', 'man', 'to',
'miserable', 'became', 'swan', 'there', 'hunt', 'show', 'i',
'heaven', 'as', 'selfish', 'after', 'suppose', 'you', 'only',
'done', 'drinking', 'then', 'care', 'it', 'him', 'come', 'swan's',
'if', 'water', 'himself', 'nothing', '"please', 'carried', 'king',
'help', 'heard', 'up', 'try', 'a', 'held', 'saw', 'life',
'surprised', 'go', '"i', 'for', 'doing', 'our', 'some', '"now',
'sometime', 'forest', 'lazy', 'not', '"you', 'replied', 'quite',
'duties', 'thought', 'one', 'and', 'went', 'his'}
```

以上是包含非重复词汇的 set 集合，词汇来自于这两篇短文。

7. 现在，最后一步就是找出两篇短文之间的共有词汇。

```
common_vocab = story1_vocab & story2_vocab
print("Common Vocabulary :",common_vocab)
```

8. Python 允许使用集合操作符 &，我们用它来找出两个词汇集中共有的词汇。最终输出结果如下：

```
Common Vocabulary : {'', 'king', 'am', 'having', 'he', 'all',
'your', 'in', 'was', 'me', 'a', 'to', 'came', 'that', 'lived',
'soon', 'saw', 'of', 'by', 'on', 'day', 'no', 'never', 'kingdom',
'there', 'for', 'i', 'said', 'will', 'the', 'this', 'as', 'angry',
'you', 'not', 'but', 'would', 'golden', 'thought', 'time', 'one',
'and', 'lake', 'went', 'water', 'his'}
```

这就是最终目标。为了节省空间，我们不再给出整个程序的输出结果。

3.7.3 工作原理

通过本节我们学会了从两篇短文中提取共有词汇的方法。我们没有使用任何其他的库也没有执行任何复杂的操作。本节也奠定了一个基础，我们创建的词汇集可以用于后续的各种任务中。

从现在开始，我们能够实现很多不同的应用任务，例如文本相似度计算、搜索引擎标注、文本摘要等。

CHAPTER 4

第 4 章 正则表达式

4.1 引言

在之前的章节中，我们学习了如何对原始数据集进行预处理。接下来我们将在本章学习如何书写和使用正则表达式（Regular Expressions）。正则表达式在我们将要学习的知识中是最简单、最基本，也是最重要、最强大的工具之一。正则表达式通常用于文本模式匹配。在本章的学习过程中，我们会了解到正则表达式的真正强大之处。

本章的目的并不是使读者成为编写正则表达式的专家。本章旨在向读者们介绍模式匹配的概念，并以这种方式进行后续的文本分析任务，其中，正则表达式是完成这些任务的最好入门工具。当读者们完成本章实例的学习后，将会对文本匹配、文本分割、文本搜索以及文本提取任务感到胸有成竹。下面正式开始我们的学习。

4.2 正则表达式——学习使用 *、+ 和?

我们用一个实例来详细说明 *、+ 和? 运算符在正则表达式中的使用方法。这些短运算符通常被称作通配符，但在本书中，为了更直观，我们称它们为零个或多个 (*)、一个或多个 (+) 和零个或一个 (?)。

4.2.1 准备工作

正则表达式的相关库是 Python 包的一部分，因此不需要安装额外的包。

4.2.2 如何实现

1. 创建一个名为 regex1.py 的文件并添加下面的导入语句：

```
import re
```

上述操作导入了 re 对象用于处理和执行正则表达式。

2. 将下面的 Python 函数添加到 regex1.py 文件中，这个函数用于匹配给定模式：

```
def text_match(text, patterns):
```

这个函数有两个输入参数，参数 text 是输入文本，参数 patterns 是对 text 内容进行匹配的模式。

3. 现在我们来定义上述函数，将以下语句添加到函数体中：

```
if re.search(patterns,  text):
  return 'Found a match!'
else:
  return('Not matched!')
```

re.search() 方法将输入模式应用于对象 text，并根据运行结果返回 true 或 false。至此，函数定义完成。

4. 接下来我们将依次使用各个通配符作为 text_match 函数的输入并观察运行结果。首先从零个或一个 (?) 开始：

```
print(text_match("ac", "ab?"))
print(text_match("abc", "ab?"))
print(text_match("abbc", "ab?"))
```

5. 注意输入模式 ab ？ 。它代表 a 后跟随零个或一个 b。相应运行结果为：

```
Found a match!
Found a match!
Found a match!
```

所有的输入文本都找到了一个匹配项。这是因为这个模式只是匹配输入的一部分，而不是整个输入。因此，三个输入都有相应的匹配项。

6. 接着测试零个或多个 (*)。在函数体内添加如下语句：

```
print(text_match("ac", "ab*"))
print(text_match("abc", "ab*"))
print(text_match("abbc", "ab*"))
```

7. 这里采用同样的输入，模式 ab* 代表 a 后跟随零个或多个 b。输出结果如下：

```
Found a match!
Found a match!
Found a match!
```

可以看到，所有输入文本都有一个匹配项。根据经验不难看出，零个或一个也必然满足零个或多个，即匹配通配符？的文本是匹配通配符 * 的文本的子集。

8. 接下来，测试一个或多个通配符 (+)。添加如下代码：

```
print(text_match("ac", "ab+"))
print(text_match("abc", "ab+"))
print(text_match("abbc", "ab+"))
```

9. 依然采用同样的输入字符串，只是改成模式 ab+，它代表 a 后跟随一个或多个 b。输出结果如下：

```
Not matched!
Found a match!
Found a match!
```

结果表明，第一个输入文本未找到匹配项，其余输入都有相应匹配项。

10. 对匹配指定重复次数，代码如下：

```
print(text_match("abbc", "ab{2}"))
```

模式 ab{2} 代表 a 后跟随两个 b。显然，在输入文本中存在一个匹配项。

11. 设定匹配重复次数的变化范围，代码如下：

```
print(text_match("aabbbbc", "ab{3,5}?"))
```

输入文本仍有匹配项，因为输入文本有一个 a 后面跟着四个 b 的子字符串 abbbb。由于我们已经分析了每步的输出结果，因此程序的整体输出结果不再赘述。

4.2.3 工作原理

re.search() 函数将输入模式 pattern 应用到输入文本 text 中，并根据运行结果返回 true 或 false。由于该函数不会返回文本的匹配项，因此我们将在接下来的实例中学习其他的 re 函数。

4.3 正则表达式——学习使用 $ 和 ^，以及如何在单词内部（非开头与结尾处）进行模式匹配

以（^）开始并以（$）操作符结尾的模式通常用于匹配一个输入文本的开始或结尾处。

4.3.1 准备工作

我们可以重用上一节中的 text_match() 函数，但相对于导入外部文件，更好的方式是重写它。接下来我们来学习如何实现本节实例。

4.3.2 如何实现

1. 创建一个名为 regex2.py 的文件并添加下面的导入语句：

```
import re
```

2. 将下面的 Python 函数添加到创建的新文件中，这个函数用于匹配给定模式：

```
def text_match(text, patterns):
  if re.search(patterns,  text):
    return 'Found a match!'
  else:
    return('Not matched!')
```

该函数有两个输入参数，参数 text 是输入文本，参数 patterns 是对 text 文本内容进行匹配的模式。函数返回结果为是否有相应匹配项。不难看出，这个函数与我们在上个实例中使用的函数完全相同。

3. 我们先来学习一个简单例子，应用下面的模式串：

```
print("Pattern to test starts and ends with")
print(text_match("abbc", "^a.*c$"))
```

4. 注意输入的模式 ^a.*c$，它代表：以 a 开头，中间有零个或多个任意字符，并以 c 结尾。上述语句的执行结果为：

```
Pattern to test starts and ends with

Found a match!
```

显然，函数为输入文本找到了一个匹配项。在这里，我们要学习一个新的通配符："."。在默认模式下，"."可以匹配除换行符以外的所有字符，即当你输入".*"时，意味着出现零个或多个任意字符。

5. 接下来我们通过模式匹配来查找以某个单词开头的文本，添加如下两行：

```
print("Begin with a word")
print(text_match("Tuffy eats pie, Loki eats peas!", "^\w+"))
```

6. 其中，\w 代表字母、数字以及下划线。该模式代表：从文本开始符号（^）开始，以任意字母或数字符号（\w）出现一次或多次（+）的文本。输出结果如下：

```
Begin with a word

Found a match!
```

果不其然，输入模式找到了匹配项。

7. 接下来，我们检查输入文本是否以单词和任意标点为结尾。添加以下几行代码：

```
print("End with a word and optional punctuation")
print(text_match("Tuffy eats pie, Loki eats peas!", "\w+\S*?$"))
```

8. 该模式代表输入文本的末尾有一个或多个 \w 出现，且其后是零个或多个 \S，并且最后匹配到文本的结尾处。为了便于理解 \S(大写字母 S)，我们必须先了解 \s 代表空格字符。\S 代表的集合正好与 \s 代表的集合相反，当它跟在符号 \w 后面意味着 \w 后有一个标点符号：

```
End with a word and optional punctuation

Found a match!
```

显然，在文本末尾存在一个匹配项。

9. 接下来，要找出包含一个特定字符的单词，添加以下几行代码：

```
print("Finding a word which contains character, not start or end of
the word")
print(text_match("Tuffy eats pie, Loki eats peas!", "\Bu\B"))
```

同理，\B 是 \b 的反向或反向集合。\b 代表单词的开头和结尾处的空字符串，由此我们可以识别出单词的边界。因此，\B 可以匹配单词内部，这里它会匹配输入文本中任何包含字符 u 的单词：

```
Finding a word which contains character, not start or end of the
word

Found a match!
```

匹配项为输入文本的第一个单词：Tuffy。

程序的整体输出如下。每个输出结果都已在上面详细讨论过，这里不再赘述：

```
Pattern to test starts and ends with

Found a match!

Begin with a word

Found a match!

End with a word and optional punctuation

Found a match!

Finding a word which contains character, not start or end of the word

Found a match!
```

4.3.3 工作原理

本节我们学习了如何在开头和结尾处进行模式匹配，还学习了通配符 . 以及其他一些特殊的字符，比如 \w、\s、\b 等。

4.4 匹配多个字符串和子字符串

在本节中，我们将使用正则表达式运行一些迭代函数。更具体地说，我们将使用 for 循环在一个输入文本上运行多个模式，并且还可以在输入文本上使用单个模式实现多次匹配。下面我们开始本节的学习。

4.4.1 准备工作

下载并安装 PyCharm 或其他你常用的 Python 编辑器。

4.4.2 如何实现

1. 创建一个名为 regex3.py 的文件并添加下面的导入语句：

```
import re
```

2. 以下两行代码为定义模式和输入文本：

```
patterns = [ 'Tuffy', 'Pie', 'Loki' ]
text = 'Tuffy eats pie, Loki eats peas!'
```

3. 添加如下循环语句：

```
for pattern in patterns:
  print('Searching for "%s" in "%s" -&gt;' % (pattern, text),)
  if re.search(pattern,  text):
    print('Found!')
  else:
    print('Not Found!')
```

这是一个简单的 for 循环，它会逐个迭代模式列表 patterns 中的项并调用 re.search() 函数。运行程序后，我们会发现 patterns 列表中的三个模式有两个会在输入文本中找到匹配项。另外，请注意，以上模式匹配是区分大小写的。我们将在后面的输出部分详细讨论这个结果。

4. 接下来我们要在输入文本中找到一个子字符串并确定它的位置。相应的输入文本和模式定义如下：

```
text = 'Diwali is a festival of lights, Holi is a festival of
colors!'
pattern = 'festival'
```

这两行分别定义了输入文本和要搜索的模式。

5. 现在，for 循环将在输入文本上遍历并获取给定模式的所有匹配项：

```
for match in re.finditer(pattern, text):
  s = match.start()
  e = match.end()
  print('Found "%s" at %d:%d' % (text[s:e], s, e))
```

6. finditer 函数将模式和输入文本作为输入，并对返回列表进行迭代。针对列表中每一个对象，都会调用 match.start() 和 match.end() 方法以找到输入文本中匹配项的确切位置。输出结果如下所示：

```
Found "festival" at 12:20

Found "festival" at 42:50
```

输出结果有两行。这表明我们在输入文本的两处找到了模式匹配项。第一个在位置 12:20，第二个在 42:50，如上所示。

程序的整体输出如下。我们已详细讨论了部分输出结果，接下来我们将对整体结果进行分析：

```
Searching for "Tuffy" in "Tuffy eats pie, Loki eats peas!" -&gt;

Found!

Searching for "Pie" in "Tuffy eats pie, Loki eats peas!" -&gt;
```

```
Not Found!

Searching for "Loki" in "Tuffy eats pie, Loki eats peas!" -&gt;

Found!

Found "festival" at 12:20

Found "festival" at 42:50
```

前六行的输出结果很直观。程序找到了 Tuffy 和 Loki 的匹配项，未找到 Pie 的匹配项（因为 re.search() 函数区分大小写）。我们已经在第六步中详细讨论过了最后两行代码，故在此不再赘述。程序的整体功能为：在输入文本中搜索字符串并给出相应匹配项的索引位置。

4.4.3　工作原理

在此我们探讨下目前频繁使用的 re.search() 函数。从先前的输出结果中可以看出，单词 pie 是输入文本的一部分，但当程序搜索大写单词 Pie 时并没有定位到它。如果要求搜索函数不区分大小写，可以定义一个标志 re.IGNORECASE。调整后的函数为 re.search(pattern，string，flags = re.IGNORECASE)。

re.finditer() 函数的用法是 re.finditer(pattern,string,flags=0)。它返回一个包含 MatchObject 对象的迭代器，可以获取输入字符串的所有非重叠匹配。

4.5　学习创建日期正则表达式和一组字符集合或字符范围

在本节中，我们先从简单的日期正则表达式入手，随后我们将学习括号 () 的用法并进一步学习如何使用方括号 []。其中，方括号 [] 表示一个集合（我们会详细介绍集合的定义）。

4.5.1　如何实现

1. 创建一个名为 regex4.py 的文件并添加下面的导入语句：

```
import re
```

2. 声明一个 url 对象，并创建一个用于查找日期的正则表达式：

```
url=
"http://www.telegraph.co.uk/formula-1/2017/10/28/mexican-grand-prix
-2017-time-does-start-tv-channel-odds-lewis1/"

date_regex = '/(\d{4})/(\d{1,2})/(\d{1,2})/'
```

url 是一个简单的字符串对象。date_regex 也是一个简单的字符串对象，但它包含了一个正则表达式，它将匹配格式为 YYYY / DD / MM 或 YYYY / MM / DD 的日期。\d 表示数字 0 到 9。在先前章节中我们已经学习过符号 {}。

3. 使用 date_regex 匹配 url，添加以下代码：

```
print("Date found in the URL :", re.findall(date_regex, url))
```

4. 函数 re.findall(pattern，input，flags = 0) 是一个新函数，它的输入参数为模式、输入文本和可选标志（先前实例中区分大小写的标志）。输出结果如下：

```
Date found in the URL : [('2017', '10', '28')]
```

显然，程序在输入文本中找到了匹配日期：2017 年 10 月 28 日。

5. 接下来，我们将学习如何使用字符集合符号 []。添加以下函数：

```
def is_allowed_specific_char(string):
  charRe = re.compile(r'[^a-zA-Z0-9.]')
  string = charRe.search(string)
  return not bool(string)
```

该函数的功能为检查输入字符串是否包含一组特定的字符或其他字符。与之前函数的不同之处在于：首先，我们对给定模式调用 re.compile() 函数，该函数返回一个类型为 RegexObject 的对象。然后，我们在已编译的模式上调用 RegexObject 的 search 方法。如果找到匹配项，search 方法将返回一个类型为 MatchObject 的对象，否则返回 None。接下来我们来看集合符号 []，方括号内的模式代表：非在字符 a-z、A-Z、0-9 或 . 的范围内，其中 ^ 代表非。

实际上，方括号内的所有内容都是或（OR）关系。

6. 使用两个不同的输入来测试上述模式，一个有匹配项，一个没有：

```
print(is_allowed_specific_char("ABCDEFabcdef123450."))
print(is_allowed_specific_char("*&%@#!}{"))
```

7. 第一组输入字符包含所有合法的字符列表，而第二组包含所有不合法的字符集合。输出结果如下：

```
True

False
```

该模式将遍历输入字符串的每个字符，查看是否有任何不合法的字符，若有则将其标记出来。读者们可以尝试在第一次调用 is_allwoed_specific_char() 时添加任何不合法的字符集，并自行检查输出结果。

程序的整体输出如下。每步输出结果都已在上面详细讨论过，这里不再赘述：

```
Date found in the URL : [('2017', '10', '28')]

True

False
```

4.5.2 工作原理

在此我们先讨论组的定义。正则表达式中的组表示模式声明中括号 () 所包含的内容。如果读者注意观察日期匹配的输出，将会发现一个集合符号 []，其中包含三个字符串对象：[('2017'，'10'，'28')]。现在仔细观察声明的模式：/(\d {4})/(\d {1,2})/(\d {1,2})/。日期的三个组成部分都在组的符号 () 内声明，且三个部分都是独立声明的。

re.findall() 方法将查找给定输入的所有匹配项。这意味着如果给定的输入文本有多个日期，那么会有多个输出结果，例如：[('2017'，'10'，'28')，('2015'，'05'，'12')]。

集合符号 [] 实质上用于匹配集合符号内包含的任一字符。只要发现匹配项，则匹配该模式成功。

4.6 查找句子中所有长度为 5 的单词，并进行缩写

至此，我们已经学习了所有的重要通配符并在实例中展示了它们各自的用法。接下来我们将在实例中进一步学习如何使用正则表达式来完成某个具体的任务。当然，在本节我们还会学习到更多的符号。

4.6.1 如何实现

1. 创建一个名为 regex_assignment1.py 的文件并添加下面的导入语句：

```
import re
```

2. 定义输入文本，使用替换功能来实现缩写，语句如下：

```
street = '21 Ramkrishna Road'
print(re.sub('Road', 'Rd', street))
```

3. 使用 re.sub() 方法来实现缩写。我们的任务是首先查找字符串 Road，然后将其替换为字符串 Rd。输入文本是字符串对象 street。输出结果如下：

```
21 Ramkrishna Rd
```

显然，输出与预期结果一致。

4. 接下来我们学习如何在任意给定的句子中找到长度为 5 的单词。添加以下语句：

```
text = 'Diwali is a festival of light, Holi is a festival of
color!'
print(re.findall(r"\b\w{5}\b", text))
```

5. 首先声明字符串对象 text 并为其赋值，接着创建一个模式并调用 re.findall() 函数。使用 \b 边界集来确定单词的边界，并使用 {} 符号以确保我们只匹配长度为 5 的单词。运行结果如下所示：

```
['light', 'color']
```

程序的整体输出如下。每步输出结果都已在上面详细讨论过，因此这里不再赘述：

```
21 Ramkrishna Rd

['light', 'color']
```

4.6.2　工作原理

至此，通过以上实例的学习，相信读者们对正则表达式的符号和语法有了较多的理解。接下来，我们探讨一些更有趣的东西。注意观察 findall() 函数，我们会看到像 r " pattern" 这样的符号，这被称为原始字符串符号。它有助于正则表达式保持条理清晰，否则就必须为正则表达式中的所有反斜杠提供转义序列。例如，模式 r"\ b \ w {5} \ b" 和 "\\ b \\ w {5} \\ b" 的功能完全相同。

4.7　学习编写基于正则表达式的分词器

在之前的章节中，我们学习了分词和分词器的概念以及它们的用途。随后，我们进一步学习了如何使用 NLTK 模块的内置分词器。在本节中，我们将学习如何自行编写分词器。它模仿了 nltk.word_tokenize() 函数的功能。

4.7.1　准备工作

下载并安装 Python 编辑器。

4.7.2　如何实现

1. 创建一个名为 regex_tokenizer.py 的文件并添加下面的导入语句：

```
import re
```

2. 对原始句子初始化，并用相应模式进行分词：

```
raw = "I am big! It's the pictures that got small."

print(re.split(r' +', raw))
```

3. 该模式和前面章节中我们学习的 space tokenizer 具有相同的功能。输出如下：

```
['I', 'am', 'big!', "It's", 'the', 'pictures', 'that', 'got',
'small.']
```

正如我们所看到的，结果与预期完全一致。

4. 当然，我们预期的结果不仅如此。我们还想用非单词的任意字符进行分割，而不仅是空格（' '）字符。尝试以下模式：

```
print(re.split(r'\W+', raw))
```

5. 使用非单词字符 \W 进行切分。结果如下：

```
['I', 'am', 'big', 'It', 's', 'the', 'pictures', 'that', 'got',
'small', '']
```

我们可以用所有的非单词字符（' ', ',', '!' 等）进行切分，但是输出结果并未保留下来这些字符。因此接下来我们要使用别的解决方法。

6. 由于 re.split() 实现不了我们想要的功能，因此我们调用另一个函数 re.findall()。添加以下行：

```
print(re.findall(r'\w+|\S\w*', raw))
```

7. 运行结果如下：

```
['I', 'am', 'big', '!', 'It', "'s", 'the', 'pictures', 'that',
'got', 'small', '.']
```

至此，我们自己的分词器已编写完成。

程序整体输出如下。每步输出结果都已在上面详细讨论过：

```
['I', 'am', 'big!', "It's", 'the', 'pictures', 'that', 'got', 'small.']

['I', 'am', 'big', 'It', 's', 'the', 'pictures', 'that', 'got', 'small',
'']

['I', 'am', 'big', '!', 'It', "'s", 'the', 'pictures', 'that', 'got',
'small', '.']
```

如上所示，我们逐渐改善了我们的模式和方法，最终达到了预期结果。

4.7.3 工作原理

我们首先调用 re.split() 采用空格字符进行分割，之后使用非单词字符替代空格字符进行分割。最后，我们改变了策略，不再通过非单词字符来分割句子，而是调用 re.findall() 来实现我们想要的功能。

4.8 学习编写基于正则表达式的词干提取器

在前面的章节中，我们学习了词干（stems/lemmas）和词干提取器（stemmer）的概念以及它们各自的用途。并且我们也已经学习了如何使用 NLTK 模块内置的波特词干提取器（porter stemmer）和兰开斯特词干提取器（lancaster stemmer）。在本节中，我们将学习如何自行编写基于正则表达式的词干提取器，它将去除不需要的后缀并找到正确的词干。

4.8.1 准备工作

和之前的词干提取器实例一样，在编写词干提取器之前，我们需要对文本进行分词处理。我们将重用上一节中最后一个分词模式对文本进行分词。

4.8.2 如何实现

1. 创建一个名为 regex_stemmer.py 的文件并添加下面的导入语句：

```
import re
```

2. 声明一个用于词干提取的函数：

```
def stem(word):
```

此函数的输入参数为一个字符串对象，返回结果为另一个字符串对象，即词干。

3. 定义该函数：

```
splits = re.findall(r'^(.*?)(ing|ly|ed|ious|ies|ive|es|s|ment)?$',
word)
stem = splits[0]
return stem
```

我们对输入的单词调用 re.findall() 函数，该函数返回两组数据。第一组数据为词干，第二组是词干任何可能的后缀。函数调用后仅返回第一组数据作为结果。

4. 定义输入文本并作分词处理。添加以下代码：

```
raw = "Keep your friends close, but your enemies closer."
tokens = re.findall(r'\w+|\S\w*', raw)
print(tokens)
```

5. 运行结果如下：

```
['Keep', 'your', 'friends', 'close', ',', 'but', 'your', 'enemies',
'closer', '.']
```

6. 对生成的单词列表逐个调用 stem() 函数。添加如下 for 循环：

```
for t in tokens:
  print("'"+stem(t)+"'")
```

以上语句循环遍历所有单词，并逐一输出词干。输出结果将在后面部分详细讨论。

程序的整体输出如下：

```
['Keep', 'your', 'friends', 'close', ',', 'but', 'your', 'enemies',
'closer', '.']

'Keep'

'your'

'friend'

'close'

','

'but'
```

```
'your'

'enem'

'closer'

'.'
```

由于词干提取器的输入句子比较简单，我们的词干提取器表现得相当不错。所以读者如果感兴趣，可以尝试对其他句子进行测试，观察效果如何。

4.8.3 工作原理

同样，我们调用 re.findall() 函数来获得期望输出。如果读者细心观察我们的第一组正则表达式，可以发现我们使用了一个非贪婪的通配符匹配（.*?)，否则，函数会贪婪地匹配到整个单词，而无法识别出后缀。此外，我们对输入句子的开始到结尾进行整个输入单词的强制性匹配，并进行切分。

CHAPTER 5

第 5 章

词性标注和文法

5.1 引言

本章主要以 Python NLTK 为工具学习以下主题：

- 词性标注器（Tagger）
- 上下文无关文法（CFG）

词性标注是对给定句子中的单词进行**词性**（Parts of Speech，POS）分类的过程。实现标注目的的软件称为**词性标注器**（tagger）。NLTK 支持多种标注器。本章的部分内容将介绍以下标注器：

- 内置标注器
- 默认标注器
- 正则表达式标注器
- 查询标注器

上下文无关文法描述了一组可以应用于构造正规语言的文本规则集合，并能以此生成新的文本集合。

语言中的上下文无关文法包括以下内容：

- 开始符号
- 终结符号（结束符号）集合
- 非终结符号（非结束符号）集合
- 用于将非终结符号转换成终结符号或非终结符号的重写规则或产生式

5.2 使用内置的词性标注器

在下面的示例中，我们使用 Python NLTK 库在给定的文本中了解更多关于词性的特征。

我们将使用 Python NLTK 库中的以下技术：

- Punkt 英文分词器（Punkt English tokenizer）
- 平均感知机标注器（Averaged perception tagger）

可以通过 Python 提示符调用 nltk.download() 函数，从 NLTK 下载这些标注器的数据集。

5.2.1　准备工作

你应该在系统中安装可用的 Python（首选 Python 3.6 版本）以及 NLTK 库及其所有的集合以获得最佳体验。

5.2.2　如何实现

1. 打开 Atom 编辑器（或者你常用的程序编辑器）。
2. 创建一个新文件，命名为 Exploring.py。
3. 输入以下源代码：

```
Exploring.py
1 import nltk
2 simpleSentence = "Bangalore is the capital of Karnataka."
3 wordsInSentence = nltk.word_tokenize(simpleSentence)
4 print(wordsInSentence)
5 partsOfSpeechTags = nltk.pos_tag(wordsInSentence)
6 print(partsOfSpeechTags)
7
```

4. 保存文件。
5. 使用 Python 编译器运行程序。
6. 你将看到如下输出：

```
ch5 — -bash — 98×21
nltk $ python Exploring.py
['Bangalore', 'is', 'the', 'capital', 'of', 'Karnataka', '.']
[('Bangalore', 'NNP'), ('is', 'VBZ'), ('the', 'DT'), ('capital', 'NN'), ('of', 'IN'), ('Karnataka'
, 'NNP'), ('.', '.')]
nltk $
```

5.2.3　工作原理

现在，我们来看一下我们刚才编写的程序，并深入研究细节：

```
import nltk
```

这是我们程序中的第一条代码，它指示 Python 解释器将 NLTK 模块从磁盘加载到内存，并使 NLTK 库在程序中可用。

```
simpleSentence = "Bangalore is the capital of Karnataka."
```

通过这条代码，我们创建了一个名为 simpleSentence 的变量，并定义为一个字符串。

```
wordsInSentence = nltk.word_tokenize(simpleSentence)
```

通过这条代码，我们调用了 NLTK 内置的 word_tokenize() 函数，它将给定的句子分解成单词并返回一个 Python 列表数据类型。一旦得到由函数计算的结果，我们使用“=(equal to)”运算符将其分配给 wordsInSentence 变量。

```
print(wordsInSentence)
```

通过这条代码，我们调用了 Python 内置的 print() 函数，它显示了如屏幕所示的数据结构。在本例中，我们显示了所有标注的单词列表。仔细观察这些输出，屏幕上显示了一个 Python 列表的数据结构，它由以逗号分隔的所有字符串组成，所有的列表元素都是括在方括号内。

```
partsOfSpeechTags = nltk.pos_tag(wordsInSentence)
```

通过这条代码，我们调用了内置的标注器 pos_tag()，通过 wordsInsentence 变量生成一个单词列表，同时也标注了词性。一旦标注完成，就会生成完整的元组列表，每个元组都有标记的单词和相应的词性标注。

```
print(partsOfSpeechTags)
```

通过这条代码，我们调用 Python 内置的 print() 函数，它将在屏幕上打印出给定的参数。在这个实例中，我们可以看到元组列表中的每个元组都由原始单词和 POS 标注组成。

5.3 编写你的词性标注器

在下面的实例中，我们使用 NLTK 库来编写你自己的词性标注器，将涉及以下类型的词性标注器：

- 默认标注器
- 正则表达式标注器
- 查询标注器

5.3.1 准备工作

首先你的系统中需要安装一个可用的 Python（首选 Python 3.6）以及 NLTK 库及其所有的集合以获得最佳体验。

你也应该安装 python-crfsuite 来运行这个实例。

5.3.2 如何实现

1. 打开 Atom 编辑器（或者你常用的程序编辑器）。

2. 创建一个新文件，命名为 OwnTagger.py。

3. 输入以下源代码：

```
OwnTagger.py
import nltk
def learnDefaultTagger(simpleSentence):
    wordsInSentence = nltk.word_tokenize(simpleSentence)
    tagger = nltk.DefaultTagger("NN")
    posEnabledTags = tagger.tag(wordsInSentence)
    print(posEnabledTags)
def learnRETagger(simpleSentence):
    customPatterns = [
        (r'.*ing$', 'ADJECTIVE'),                   # running
        (r'.*ly$', 'ADVERB'),                       # willingly
        (r'.*ion$', 'NOUN'),                        # intimation
        (r'(.*ate|.*en|is)$', 'VERB'),              # terminate, darken, lighten
        (r'^an$', 'INDEFINITE-ARTICLE'),            # terminate
        (r'^(with|on|at)$', 'PREPOSITION'),         # on
        (r'^\-?[0-9]+(\.[0-9]+)$', 'NUMBER'),       # -1.0, 12345.123
        (r'.*$', None),
    ]
    tagger = nltk.RegexpTagger(customPatterns)
    wordsInSentence = nltk.word_tokenize(simpleSentence)
    posEnabledTags = tagger.tag(wordsInSentence)
    print(posEnabledTags)
def learnLookupTagger(simpleSentence):
    mapping = {
        '.': '.', 'place': 'NN', 'on': 'IN',
        'earth': 'NN', 'Mysore' : 'NNP', 'is': 'VBZ',
        'an': 'DT', 'amazing': 'JJ'
    }
    tagger = nltk.UnigramTagger(model=mapping)
    wordsInSentence = nltk.word_tokenize(simpleSentence)
    posEnabledTags = tagger.tag(wordsInSentence)
    print(posEnabledTags)

if __name__ == '__main__':
    testSentence = "Mysore is an amazing place on earth. I have visited Mysore 10 times."
    learnDefaultTagger(testSentence)
    learnRETagger(testSentence)
    learnLookupTagger(testSentence)
```

4. 保存文件。

5. 使用 Python 编译器运行程序。

6. 你将看到如下输出：

```
ch5 — -bash — 98×21
nltk $ python OwnTagger.py
[('Mysore', 'NN'), ('is', 'NN'), ('an', 'NN'), ('amazing', 'NN'), ('place', 'NN'), ('on', 'NN'), (
'earth', 'NN'), ('.', 'NN'), ('I', 'NN'), ('have', 'NN'), ('visited', 'NN'), ('Mysore', 'NN'), ('1
0', 'NN'), ('times', 'NN'), ('.', 'NN')]
[('Mysore', None), ('is', 'VERB'), ('an', 'INDEFINITE-ARTICLE'), ('amazing', 'ADJECTIVE'), ('place
', None), ('on', 'PREPOSITION'), ('earth', None), ('.', None), ('I', None), ('have', None), ('visi
ted', None), ('Mysore', None), ('10', None), ('times', None), ('.', None)]
[('Mysore', 'NNP'), ('is', 'VBZ'), ('an', 'DT'), ('amazing', 'JJ'), ('place', 'NN'), ('on', 'IN'),
 ('earth', 'NN'), ('.', '.'), ('I', None), ('have', None), ('visited', None), ('Mysore', 'NNP'), (
'10', None), ('times', None), ('.', '.')]
nltk $
```

5.3.3 工作原理

现在，我们来进一步了解我们刚编写的程序：

```
import nltk
```

这是我们程序中的第一条代码，它指示 Python 解释器将 NLTK 模块从磁盘加载到内存，使得 NLTK 库在程序中可用：

```
def learnDefaultTagger(simpleSentence):
  wordsInSentence = nltk.word_tokenize(simpleSentence)
  tagger = nltk.DefaultTagger("NN")
  posEnabledTags = tagger.tag(wordsInSentence)
  print(posEnabledTags)
```

所有的这些代码定义了一个新的 Python 函数，它的输入是一个字符串，然后在屏幕上打印出句子中的单词及默认标注。我们来进一步理解这个函数的功能：

```
def learnDefaultTagger(simpleSentence):
```

通过这条代码，我们定义了一个新的 Python 函数，名为 learnDefaultTagger，它接收一个名为 simpleSentence 的参数。

```
wordsInSentence = nltk.word_tokenize(simpleSentence)
```

通过这条代码，我们从 NLTK 库调用 word_tokenize 函数。将 simpleSentence 作为第一个参数传递给这个函数。调用这个函数处理数据后，返回值是一个单词列表，并存储在 InSentence 变量中。

```
tagger = nltk.DefaultTagger("NN")
```

通过这条代码，我们在 Python NLTK 库中将参数 NN 传递给 DefaultTagger() 类来创建一个 tagger 对象。

```
posEnabledTags = tagger.tag(wordsInSentence)
```

通过这条代码，我们调用了 tagger 对象的 tag() 函数。从 wordsInSentence 变量中得到了单词序列，然后返回了标注后的单词列表，并保存在 posEnabledTags 中。值得注意的是，句子中的所有单词都标注为 NN，这是标注器之前设置好的，用于不知道词性信息时的基本标注。

```
print(posEnabledTags)
```

这里，我们调用 Python 内置的 print() 函数来检查 posEnabledTags 变量的内容。我们可以看到句子中的所有单词都会被标注为 NN。

```
def learnRETagger(simpleSentence):
  customPatterns = [
    (r'.*ing$', 'ADJECTIVE'),
    (r'.*ly$', 'ADVERB'),
    (r'.*ion$', 'NOUN'),
```

```
    (r'(.*ate|.*en|is)$', 'VERB'),
    (r'^an$', 'INDEFINITE-ARTICLE'),
    (r'^(with|on|at)$', 'PREPOSITION'),
    (r'^\-?[0-9]+(\.[0-9]+)$', 'NUMBER'),
    (r'.*$', None),
  ]
  tagger = nltk.RegexpTagger(customPatterns)
  wordsInSentence = nltk.word_tokenize(simpleSentence)
  posEnabledTags = tagger.tag(wordsInSentence)
  print(posEnabledTags)
```

这些代码创建了一个名为 learnRETagger() 的新函数，它的输入是一个字符串，输出是这个字符串包含的所有单词的列表，以及根据正则表达式标注器识别出的词性。

我们尝试每一次只理解一条代码：

```
def learnRETagger(simpleSentence):
```

我们正在定义一个新的 Python 函数，名为 learnRETagger，参数为 simpleSentence。

为了理解下一条代码，我们应该学习更多关于 Python 列表、元组和正则表达式的内容：

- Python 列表是一种数据结构，它包含一组有序的元素。
- Python 元组是一种不可变的（只读）数据结构，它包含一组有序的元素。
- Python 正则表达式是以字母 r 开头的字符串，并遵循 PCRE 标注标准：

```
customPatterns = [
  (r'.*ing$', 'ADJECTIVE'),
  (r'.*ly$', 'ADVERB'),
  (r'.*ion$', 'NOUN'),
  (r'(.*ate|.*en|is)$', 'VERB'),
  (r'^an$', 'INDEFINITE-ARTICLE'),
  (r'^(with|on|at)$', 'PREPOSITION'),
  (r'^\-?[0-9]+(\.[0-9]+)$', 'NUMBER'),
  (r'.*$', None),
]
```

这些代码看起来很多，但是每条代码可以完成很多事情：

- 创建一个名为 customPatterns 的变量
- 定义一个新的 Python 列表数据类型
- 将八个元素添加到这个列表中
- 这个列表中的每个元素都是一个元组，每个元组都有两项
- 元组中的第一项是一个正则表达式
- 元组中的第二项是一个字符串

现在，把前面的这些代码翻译成人类可读的形式，我们加入了八个正则表达式将句子中的单词标注为形容词（adjective）、副词（adverb）、名词（noun）、动词（verb）、不定冠词（indefinite-article）、介词（preposition）、数词（number）或（none）类型中的任意一类。

我们通过模式的识别将英语单词标注为某一类给定的词性。

在前面的例子中，我们使用以下特征标注英文单词的词性：

- 以 ing 结尾的单词可以被标注为形容词，例如，running
- 以 ly 结尾的单词可以被标注为副词，例如，willingly
- 以 ion 结尾的单词可以被标注为名词，例如，intimation
- 以 ate 或 en 结尾的单词可以被标注为动词，例如，terminate、darken 或 lighten
- 以 an 结尾的单词可以被标注为不定冠词
- 例如 with、on 或者 at 这类词被标注为介词
- 例如 −123.0、984 这样的词可以被标注为数词
- 我们将其他单词都标注为 none，这个 Python 内置的数据类型表示无意义。

```
tagger = nltk.RegexpTagger(customPatterns)
```

通过这条代码，我们创建了 NLTK 的内置正则表达式标记器 RegexpTagger 的一个对象。我们将元组列表传递给该类的第一个参数 customPatterns 变量来初始化对象。该对象在将来可以被名为 Tagger 的变量引用。

```
wordsInSentence = nltk.word_tokenize(simpleSentence)
```

之后的处理过程，我们首先尝试使用 NLTK 内置的 word_tokenize() 函数来对字符串 simpleSentence 进行分词，并将单词列表存储在 wordsInSentence 变量中。

```
posEnabledTags = tagger.tag(wordsInSentence)
```

现在我们调用正则表达式词性标注器的 tag() 函数来标注存储在 wordsInSentence 变量中的所有单词。这个标注过程的结果存储在 posEnabledTags 变量中。

```
print(posEnabledTags)
```

我们调用 Python 内置的 print() 函数在屏幕上打印出 posEnabledTags 数据结构的内容：

```
def learnLookupTagger(simpleSentence):
  mapping = {
    '.': '.', 'place': 'NN', 'on': 'IN',
    'earth': 'NN', 'Mysore' : 'NNP', 'is': 'VBZ',
  'an': 'DT', 'amazing': 'JJ'
}
tagger = nltk.UnigramTagger(model=mapping)
wordsInSentence = nltk.word_tokenize(simpleSentence)
posEnabledTags = tagger.tag(wordsInSentence)
print(posEnabledTags)
```

仔细观察这条代码：

```
def learnLookupTagger(simpleSentence):
```

我们正在定义一个新的函数 learnLookupTagger，它将一个字符串作为参数传递给 simpleSentence 变量。

```
tagger = nltk.UnigramTagger(model=mapping)
```

通过这条代码，我们从 NLTK 库调用 UnigramTagger。这是一个查询词性标注器，它的输入是我们已经创建的定义为 mapping 变量的 Python 字典。创建的 tagger 对象用于将来调用。

```
wordsInSentence = nltk.word_tokenize(simpleSentence)
```

这里，我们使用 NLTK 内置的 word_tokenize() 函数来标注句子并将结果保存在 wordsInSentence 变量中。

```
posEnabledTags = tagger.tag(wordsInSentence)
```

当标注句子时，我们调用 tagger 的 tag() 函数，输入是传递给 wordsInSentence 变量的单词列表，并将标注的结果保存在 posEnabledTags 变量中。

```
print(posEnabledTags)
```

通过这条代码，我们将 posEnabledTags 的数据结构打印在屏幕上以便进一步检查。

```
testSentence = "Mysore is an amazing place on earth. I have visited Mysore
10 times."
```

我们正在创建一个名为 testSentence 的变量，并定义为一条简单的英语句子。

```
learnDefaultTagger(testSentence)
```

我们调用本例创建的 learnDefaultTagger() 函数，testSentence 变量作为它的第一个参数。当执行完成这个函数，我们将得到词性标注后的句子。

```
learnRETagger(testSentence)
```

通过这条代码，我们调用 learnRETagger() 函数，输入同样是保存在 testSentence 变量中的测试句子。该函数的输出是按照我们自己定义的正则表达式标注的标注列表。

```
learnLookupTagger(testSentence)
```

learnLookupTagger() 函数的输出是 testSentence 句子中所有标注的列表，它采用我们创建的查询字典进行标注。

5.4 训练你的词性标注器

在本节中，我们将学习如何训练我们自己的标注器，并将训练好的模型保存到磁盘，用于后续处理使用。

5.4.1 准备工作

你应该在你的系统中安装一个可用的 Python（首选 Python 3.6）和 NLTK 库及其所有的集合以获得最佳体验。

5.4.2　如何实现

1. 打开 Atom 编辑器（或者你常用的程序编辑器）。
2. 创建一个新文件，命名为 Train3.py。
3. 输入以下源代码：

```
Train3.py
import nltk
import pickle

def sampleData():
    return [
        "Bangalore is the capital of Karnataka.",
        "Steve Jobs was the CEO of Apple.",
        "iPhone was Invented by Apple.",
        "Books can be purchased in Market.",
    ]

def buildDictionary():
    dictionary = {}
    for sent in sampleData():
        partsOfSpeechTags = nltk.pos_tag(nltk.word_tokenize(sent))
        for tag in partsOfSpeechTags:
            value = tag[0]
            pos = tag[1]
            dictionary[value] = pos
    return dictionary

def saveMyTagger(tagger, fileName):
    fileHandle = open(fileName, "wb")
    pickle.dump(tagger, fileHandle)
    fileHandle.close()

def saveMyTraining(fileName):
    tagger = nltk.UnigramTagger(model=buildDictionary())
    saveMyTagger(tagger, fileName)

def loadMyTagger(fileName):
    return pickle.load(open(fileName, "rb"))

sentence = 'Iphone is purchased by Steve Jobs in Bangalore Market'
fileName = "myTagger.pickle"

saveMyTraining(fileName)

myTagger = loadMyTagger(fileName)

print(myTagger.tag(nltk.word_tokenize(sentence)))
```

4. 保存文件。
5. 使用 Python 编译器运行程序。
6. 你将看到如下输出：

```
ch5 — -bash — 98×21
nltk $ python Train3.py
[('Iphone', None), ('is', 'VBZ'), ('purchased', 'VBN'), ('by', 'IN'), ('Steve', 'NNP'), ('Jobs', '
NNP'), ('in', 'IN'), ('Bangalore', 'NNP'), ('Market', 'NNP')]
nltk $
```

5.4.3 工作原理

我们来了解一下程序的工作原理：

```
import nltk
import pickle
```

通过这两行代码，我们将 NLTK 和 pickle 模块加载到程序中。pickle 模块实现了强大的序列化和反序列化算法，可用于处理非常复杂的 Python 对象。

```
def sampleData():
  return [
    "Bangalore is the capital of Karnataka.",
    "Steve Jobs was the CEO of Apple.",
    "iPhone was Invented by Apple.",
    "Books can be purchased in Market.",
  ]
```

通过这些代码，我们定义了一个名为 sampleData() 的函数，它返回一个 Python 列表。本例中，我们返回四个示例字符串。

```
def buildDictionary():
  dictionary = {}
  for sent in sampleData():
    partsOfSpeechTags = nltk.pos_tag(nltk.word_tokenize(sent))
    for tag in partsOfSpeechTags:
      value = tag[0]
      pos = tag[1]
      dictionary[value] = pos
    return dictionary
```

我们定义一个名为 buildDictionary() 的函数，它每次从 sampleData() 函数生成的列表中读取一条字符串。每条字符串都用 nltk.word_tokenize() 函数进行分词。分词结果被添加到一个 Python 字典中，其中字典的键（key）是句子中的单词，字典的值（value）是词性。当字典生成完毕，返回给调用者。

```
def saveMyTagger(tagger, fileName):
  fileHandle = open(fileName, "wb")
  pickle.dump(tagger, fileHandle)
  fileHandle.close()
```

通过这些代码，我们定义一个名为 saveMyTagger() 的函数，它需要两个参数：

- `tagger`：词性标注器的一个对象
- `fileName`：存储 tagger 对象的文件名称

我们首先以**写入二进制**（Write Binary，WB）的模式打开文件。然后使用 pickle 模块的 dump() 方法，将整个标注器存储在文件中，并在 fileHandle 上调用 close() 函数。

```
def saveMyTraining(fileName):
  tagger = nltk.UnigramTagger(model=buildDictionary())
  saveMyTagger(tagger, fileName)
```

通过这些代码，我们定义一个名为 saveMyTraining 的新函数，它接收名为 fileName 的单个参数。

我们构建一个 nltk.UnigramTagger() 对象，它的输入是 buildDictionary() 函数的输出模型（根据我们之前定义的字符串示例集合建立），当生成 tagger 对象时，调用 saveMyTagger() 函数将其保存到磁盘。

```
def loadMyTagger(fileName):
  return pickle.load(open(fileName, "rb"))
```

这里，我们定义一个新的函数 loadMyTagger()，它接收名为 fileName 的单个参数。该函数从磁盘读取文件并将其传递给 pickle.load() 函数，该函数从磁盘反序列化 tagger 对象并返回它的引用。

```
sentence = 'Iphone is purchased by Steve Jobs in Bangalore Market'
fileName = "myTagger.pickle"
```

通过这两条代码，我们定义了两个变量，一个是 sentence，一个是 fileName，它们分别是我们需要分析的字符串和我们存储词性标注器的文件路径。

```
saveMyTraining(fileName)
```

这条代码实际上调用了 saveMyTraining() 函数，myTagger.pickle 为其参数。所以，我们基本上完成了将训练好的标注器存储在这个文件中。

```
myTagger = loadMyTagger(fileName)
```

通过这条代码，我们将 myTagger.pickle 作为 loadMyTagger() 函数的参数，它从磁盘加载这个标注器，对其进行反序列化并创建一个对象。该对象进一步赋给 myTagger 变量。

```
print(myTagger.tag(nltk.word_tokenize(sentence)))
```

通过这条代码，我们调用刚从磁盘加载的 tagger 对象的 tag() 函数。我们用它来标注我们创建的字符串样本。

处理完成后，屏幕上显示输出结果。

5.5 学习编写你的文法

在自动机理论中，上下文无关文法包含以下内容：

- 开始符号 / 标记

- 终结符号集合
- 非终结符号集合
- 定义开始符号和所有可能符号的规则（或产生式）

这些符号可以是我们所考虑的语言的任意特定形式。例如：

- 当语言是英文时，a、b、c、d、e、f、g、h、i、j、k、l、m、n、o、p、q、r、s、t、u、v、w、x、y、z 是符号 / 标记 / 字母。
- 当语言是十进制数字时，0、1、2、3、4、5、6、7、8、9 是符号 / 标记 / 字母。

一般来说，规则（或产生式）是用**巴克斯 – 诺尔（BNF）**范式编写的。

5.5.1　准备工作

在你的系统中安装一个可用的 Python 工具（首选 Python 3.6）和 NLTK 库。

5.5.2　如何实现

1. 打开 Atom 编辑器（或者你常用的程序编辑器）。
2. 创建一个新文件，命名为 Grammar.py。
3. 输入以下源代码：

```
Grammar.py
import nltk
import string
from nltk.parse.generate import generate
import sys

productions = [
    "ROOT -> WORD",
    "WORD -> ' '",
    "WORD -> NUMBER LETTER",
    "WORD -> LETTER NUMBER",
]

digits = list(string.digits)
for digit in digits[:4]:
    productions.append("NUMBER -> '{w}'".format(w=digit))

letters = "' | '".join(list(string.ascii_lowercase)[:4])
productions.append("LETTER -> '{w}'".format(w=letters))

grammarString = "\n".join(productions)

grammar = nltk.CFG.fromstring(grammarString)

print(grammar)

for sentence in generate(grammar, n=5, depth=5):
    palindrome = "".join(sentence).replace(" ", "")
    print("Generated Word: {}, Size : {}".format(palindrome, len(palindrome)))
```

4. 保存文件。
5. 使用 Python 编译器运行程序。

6. 你将看到如下输出：

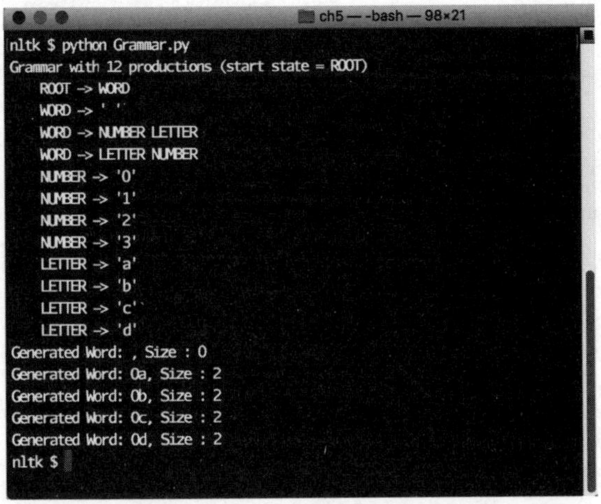

```
nltk $ python Grammar.py
Grammar with 12 productions (start state = ROOT)
    ROOT -> WORD
    WORD -> ' '
    WORD -> NUMBER LETTER
    WORD -> LETTER NUMBER
    NUMBER -> '0'
    NUMBER -> '1'
    NUMBER -> '2'
    NUMBER -> '3'
    LETTER -> 'a'
    LETTER -> 'b'
    LETTER -> 'c'
    LETTER -> 'd'
Generated Word: , Size : 0
Generated Word: 0a, Size : 2
Generated Word: 0b, Size : 2
Generated Word: 0c, Size : 2
Generated Word: 0d, Size : 2
nltk $
```

5.5.3　工作原理

现在，我们来进一步了解我们刚编写的程序：

```
import nltk
```

我们将 NLTK 库导入到当前程序中。

```
import string
```

该代码将 string 模块导入到当前程序中。

```
from nltk.parse.generate import generate
```

该代码从 nltk.parse.generate 模块导入 generate 函数，用于根据我们将要创建的上下文无关文法来生成字符串。

```
productions = [
  "ROOT -> WORD",
  "WORD -> ' '",
  "WORD -> NUMBER LETTER",
  "WORD -> LETTER NUMBER",
]
```

这里，我们定义了一个新的文法。该文法包含以下产生式规则：

- 起始符号是 ROOT
- ROOT 符号可以产生 WORD 符号
- WORD 符号可以产生 ' '（空格）。这是一条产生终结符号的产生式规则
- WORD 符号可以产生 NUMBER 符号，其后跟随 LETTER 符号
- WORD 符号可以产生 LETTER 符号，其后跟随 NUMBER 符号

以下这些代码进一步扩展了这些产生式规则：

```
digits = list(string.digits)
for digit in digits[:4]:
  productions.append("NUMBER -> '{w}'".format(w=digit))
```

- NUMBER 可以产生终结符号 0、1、2 或 3

以下代码进一步扩展了这些产生式规则：

```
letters = "' | '".join(list(string.ascii_lowercase)[:4])
productions.append("LETTER -> '{w}'".format(w=letters))
```

- LETTER 可以产生小写字母 a、b、c 或 d

我们来理解一下这个文法表示了什么语言。该文法表示包含了例如 0a、1a、2a、a1、a3 等单词的语言。

我们把当前存储在 production 列表变量中的所有产生式规则转换成一个字符串：

```
grammarString = "\n".join(productions)
```

我们使用 nltk.CFG.fromstring() 方法创建一个新的文法对象，输入是我们刚创建的 grammarString 变量：

```
grammar = nltk.CFG.fromstring(grammarString)
```

以下代码打印出了根据这个文法自动生成的语言中的前 5 个单词：

```
for sentence in generate(grammar, n=5, depth=5):
  palindrome = "".join(sentence).replace(" ", "")
  print("Generated Word: {}, Size : {}".format(palindrome,
len(palindrome)))
```

5.6 编写基于概率的上下文无关文法

基于概率的上下文无关文法是一种特殊类型的上下文无关文法，其中所有非终结符号（左侧）的概率之和应该等于 1。

我们通过一个简单的例子来加深了解。

5.6.1 准备工作

你应该在你的系统中安装一个可用的 Python（首选 Python 3.6）以及 NLTK 库及其所有的集合以获得最佳体验。

5.6.2 如何实现

1. 打开 Atom 编辑器（或者你常用的程序编辑器）。
2. 创建一个新文件，命名为 PCEG.py。
3. 输入以下源代码：

```
PCFG.py
import nltk
from nltk.parse.generate import generate

productions = [
    "ROOT -> WORD [1.0]",
    "WORD -> P1 [0.25]",
    "WORD -> P1 P2 [0.25]",
    "WORD -> P1 P2 P3 [0.25]",
    "WORD -> P1 P2 P3 P4 [0.25]",
    "P1 -> 'A' [1.0]",
    "P2 -> 'B' [0.5]",
    "P2 -> 'C' [0.5]",
    "P3 -> 'D' [0.3]",
    "P3 -> 'E' [0.3]",
    "P3 -> 'F' [0.4]",
    "P4 -> 'G' [0.9]",
    "P4 -> 'H' [0.1]",
]

grammarString = "\n".join(productions)

grammar = nltk.PCFG.fromstring(grammarString)

print(grammar)

for sentence in generate(grammar, n=10, depth=5):
    palindrome = "".join(sentence).replace(" ", "")
    print("String : {}, Size : {}".format(palindrome, len(palindrome)))
```

4. 保存文件。

5. 使用 Python 编译器运行程序。

6. 你将看到如下输出：

```
ch5 — -bash — 98×26
nltk $ python PCFG.py
Grammar with 13 productions (start state = ROOT)
    ROOT -> WORD [1.0]
    WORD -> P1 [0.25]
    WORD -> P1 P2 [0.25]
    WORD -> P1 P2 P3 [0.25]
    WORD -> P1 P2 P3 P4 [0.25]
    P1 -> 'A' [1.0]
    P2 -> 'B' [0.5]
    P2 -> 'C' [0.5]
    P3 -> 'D' [0.3]
    P3 -> 'E' [0.3]
    P3 -> 'F' [0.4]
    P4 -> 'G' [0.9]
    P4 -> 'H' [0.1]
String : A, Size : 1
String : AB, Size : 2
String : AC, Size : 2
String : ABD, Size : 3
String : ABE, Size : 3
String : ABF, Size : 3
String : ACD, Size : 3
String : ACE, Size : 3
String : ACF, Size : 3
String : ABDG, Size : 4
nltk $
```

5.6.3 工作原理

现在，我们来看看以上编写的程序：

```
import nltk
```

我们将NLTK库导入到当前的程序中。

```
from nltk.parse.generate import generate
```

该代码从nltk.parse.genearate模块导入generate函数。

```
productions = [
  "ROOT -> WORD [1.0]",
  "WORD -> P1 [0.25]",
  "WORD -> P1 P2 [0.25]",
  "WORD -> P1 P2 P3 [0.25]",
  "WORD -> P1 P2 P3 P4 [0.25]",
  "P1 -> 'A' [1.0]",
  "P2 -> 'B' [0.5]",
  "P2 -> 'C' [0.5]",
  "P3 -> 'D' [0.3]",
  "P3 -> 'E' [0.3]",
  "P3 -> 'F' [0.4]",
  "P4 -> 'G' [0.9]",
  "P4 -> 'H' [0.1]",
]
```

这里，我们为我们的语言定义文法，如下所示：

描述	内容
开始符号	ROOT
非终结符号	WORD,P1,P2,P3,P4
终结符号	'A','B','C','D','E','F','G','H'

当我们确定了文法中的符号后，产生式规则如下所示：

- 一个ROOT符号，这是该文法的开始符号
- 一个WORD符号，概率为1.0
- 一个WORD符号产生P1，概率为0.25
- 一个WORD符号产生P1 P2，概率为0.25
- 一个WORD符号产生P1 P2 P3，概率为0.25
- 一个WORD符号产生P1 P2 P3 P4，概率为0.25
- P1符号产生符号“A”，概率为1.0
- P2符号产生符号“B”，概率为0.5
- P2符号产生符号“C”，概率为0.5
- P3符号产生符号“D”，概率为0.3
- P3符号产生符号“E”，概率为0.3

- P3 符号产生符号“F”，概率为 0.4
- P4 符号产生符号“G”，概率为 0.9
- P4 符号产生符号“H”，概率为 0.1

如果仔细观察，可以发现所有非终结符号的概率之和等于 1.0。这是概率上下文无关文法的强制性要求。

我们将所有产生式规则的列表连接成一个名为 grammarString 变量的字符串：

```
grammarString = "\n".join(productions)
```

以下代码使用 nltk.PCFG.fromstring 方法创建一个 grammar 对象，它的输入是 grammarString 变量：

```
grammar = nltk.PCFG.fromstring(grammarString)
```

以下代码使用 Python 内置的 print() 函数在屏幕上打印出 grammar 对象的内容。它将统计我们刚创建的文法中的所有符号和产生式规则的总数：

```
print(grammar)
```

我们使用 NLTK 内置的 generate 函数来生成 10 个基于该文法的字符串，并在屏幕上打印出来：

```
for sentence in generate(grammar, n=10, depth=5):
  palindrome = "".join(sentence).replace(" ", "")
  print("String : {}, Size : {}".format(palindrome, len(palindrome)))
```

5.7　编写递归的上下文无关文法

递归的上下文无关文法是一类特殊类型的上下文无关文法，左侧的符号出现在产生式规则的右侧。

由递归的上下文无关文法生成的最佳范例是回文（palindromes）。对给定语言的回文，我们能编写相应的递归上下文无关文法。

我们进一步深入了解，对一个包含字母 0 和 1 的语言系统，回文表示如下：

- 11
- 1001
- 010010

无论我们从哪个方向来阅读这些字母（从左到右或从右到左），结果都是一样的。这就是回文的特点。

在本节实例中，我们将编写文法来表示这些回文，并使用 NLTK 内置的字符串生成库来生成一些回文。

我们通过一个简单的例子来加深了解。

5.7.1 准备工作

你应该在你的系统中安装一个可用的 Python（首选 Python 3.6）以及 NLTK 库。

5.7.2 如何实现

1. 打开 Atom 编辑器（或者你常用的程序编辑器）。
2. 创建一个新文件，命名为 RecursiveCFG.py。
3. 输入以下源代码：

```
RecursiveCFG.py
1  import nltk
2  import string
3  from nltk.parse.generate import generate
4
5  productions = [
6      "ROOT -> WORD",
7      "WORD -> ' '"
8  ]
9
10 alphabets = list(string.digits)
11
12 for alphabet in alphabets:
13     productions.append("WORD -> '{w}' WORD '{w}'".format(w=alphabet))
14
15 grammarString = "\n".join(productions)
16
17 grammar = nltk.CFG.fromstring(grammarString)
18
19 print(grammar)
20
21 for sentence in generate(grammar, n=5, depth=5):
22     palindrome = "".join(sentence).replace(" ", "")
23     print("Palindrome : {}, Size : {}".format(palindrome, len(palindrome)))
24
```

4. 保存文件。
5. 使用 Python 编译器运行程序。
6. 你将看到如下输出：

```
ch5 — -bash — 98×26
nltk $ python RecursiveCFG.py
Grammar with 12 productions (start state = ROOT)
    ROOT -> WORD
    WORD -> ' '
    WORD -> '0' WORD '0'
    WORD -> '1' WORD '1'
    WORD -> '2' WORD '2'
    WORD -> '3' WORD '3'
    WORD -> '4' WORD '4'
    WORD -> '5' WORD '5'
    WORD -> '6' WORD '6'
    WORD -> '7' WORD '7'
    WORD -> '8' WORD '8'
    WORD -> '9' WORD '9'
Palindrome : , Size : 0
Palindrome : 00, Size : 2
Palindrome : 0000, Size : 4
Palindrome : 0110, Size : 4
Palindrome : 0220, Size : 4
nltk $
```

5.7.3　工作原理

现在，我们来看看我们刚编写的程序，并深入细节。首先将 NLTK 库导入到我们的程序中以供将来调用：

```
import nltk
```

我们还将 string 库导入到我们的程序中以供将来调用：

```
import string
```

我们从 nltk.parse.generate 模块中导入 generate 函数：

```
from nltk.parse.generate import generate
```

我们创建了一个新的列表数据结构，名为 productions，其中包含两个元素。这两个元素都是字符串，代表上下文无关文法中的两个产生式：

```
productions = [
  "ROOT -> WORD",
  "WORD -> ' '"
]
```

我们将十进制数字以列表形式赋给 alphabets 变量：

```
alphabets = list(string.digits)
```

通过数字 0 到 9，我们添加更多的产生式到我们的列表中。以下是用来定义回文的产生式规则：

```
for alphabet in alphabets:
  productions.append("WORD -> '{w}' WORD '{w}'".format(w=alphabet))
```

当所有的规则生成后，将它们连接成一个字符串变量，名为 grammarString：

```
grammarString = "\n".join(productions)
```

以下这条代码中，我们将新构造的 grammarString 传递给 NLTK 内置的 nltk.CFG.fromstring 函数，创建了一个新的 grammar 对象：

```
grammar = nltk.CFG.fromstring(grammarString)
```

通过这条代码，我们调用 Python 内置的 print() 函数打印出我们刚创建的文法：

```
print(grammar)
```

以下我们使用 NLTK 库的 generate 函数生成 5 个回文，并在屏幕上打印出相应内容：

```
for sentence in generate(grammar, n=5, depth=5):
  palindrome = "".join(sentence).replace(" ", "")
  print("Palindrome : {}, Size : {}".format(palindrome, len(palindrome)))
```

CHAPTER 6

第6章 分块、句法分析、依存分析

6.1 引言

迄今为止，我们已经学习了如何利用 Python NLTK 对给定的文本进行词性标注。但是有时候，我们感兴趣的是如何从我们正在处理的文本中找到更多的细节。例如，在给定的文本中，我可能想要找到一些著名的人名、地名等等。我们能够将所有找到的这些名字保存到一个大字典中。其中最简单的一种方式是利用词性分析轻松地识别出这些名字。

分块（Chunking）是从文本中抽取组块短语的过程。我们将利用词性标注算法进行分块，需要注意的是，分块产生的单元或词之间不能重叠。

6.2 使用内置的分块器

在本节中，我们将学习如何使用内置的分块器。在这个过程中，我们将使用 NLTK 中的以下工具：

- Punkt 标注器（默认）
- 平均感知标注器（默认）
- 最大熵实体分块器（默认）

6.2.1 准备工作

首先，你应该已经安装好了 Python 工具包以及 NLTK 库。预先了解第 5 章的词性标注和文法也是非常有用的。

6.2.2 如何实现

1. 打开 Atom 编辑器（或者你常用的程序编辑器）。

2. 创建一个名为 Chunker.py 的新文件。

3. 输入以下源代码：

```
Chunker.py
import nltk

text = "Lalbagh Botanical Gardens is a well known botanical garden in Bengaluru, India."
sentences = nltk.sent_tokenize(text)
for sentence in sentences:
    words = nltk.word_tokenize(sentence)
    tags = nltk.pos_tag(words)
    chunks = nltk.ne_chunk(tags)
    print(chunks)
```

4. 保存文件。

5. 使用 Python 编译器运行程序。

6. 你将看到如下输出：

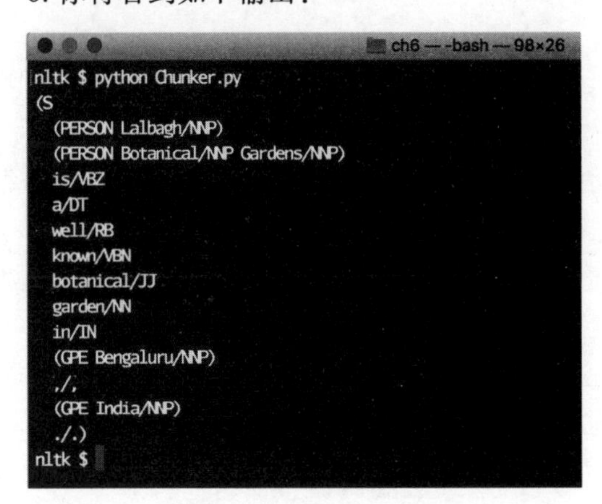

6.2.3 工作原理

现在尝试理解程序的工作原理。以下这条代码将 NLTK 模块导入程序：

```
import nltk
```

作为本例的一部分，我们将分析如下数据，将这个字符串添加到名为 text 的变量中：

```
text = "Lalbagh Botanical Gardens is a well known botanical garden in
Bengaluru, India."
```

以下代码能够将给定的文本分割成多条句子，并将分割后的句子存储在名为 sentences 变量中：

```
sentences = nltk.sent_tokenize(text)
```

以下代码循环遍历所有分割后的句子，并将每条句子存储在 sentence 变量中：

```
for sentence in sentences:
```

以下代码将句子切分成不重叠的词语，结果存储在名为 words 的变量中：

```
words = nltk.word_tokenize(sentence)
```

以下代码采用 NLTK 中的默认标注器进行词性分析，结果存储在名为 tags 的变量中：

```
tags = nltk.pos_tag(words)
```

通过以下这条代码，我们调用 nltk.ne_chunk() 的函数实现分块功能。结果存储在名为 chunks 的变量中。这个结果实际上是包含树的路径的树结构数据。

```
chunks = nltk.ne_chunk(tags)
```

以下代码打印出在给定输入字符串中识别出的组块。组块由括号“(”和“)”进行标识，这样在输入文本中能够很容易区分出组块和其他词：

```
print(chunks)
```

6.3 编写你的简单分块器

在本节中，我们将编写自己的正则表达式分块器。既然我们将使用正则表达式来编写分块器，那么就需要理解利用正则表达式进行分块的一些特点。

在第 4 章中，我们已经了解了正则表达式以及如何编写它们。例如 [a-z，A-Z]+ 这种形式的正则表达式能够匹配出句子中所有的英文单词。

我们已经熟悉了通过 NLTK 可以用简短的形式（例如 v、nn、nnp 等标签）来识别词性，那么我们能否利用这些词性来编写正则表达式呢？

答案是肯定的。我们可以利用基于词性的正则表达式进行编写。由于我们利用词性标签来编写这些正则表达式，所以称为标签模式（tag pattern）。

正如我们可以利用自然语言的字母（a-z）去匹配各种模式，同样能够根据 NLTK 识别的词性去匹配单词（在字典中的任意组合）。这些标签模式是 NLTK 最强大的特征之一，因为它们仅仅通过基于词性的正则表达式就能很灵活地匹配出句子中的单词。

为了更深入了解这些标签模式，我们进一步挖掘：

```
"Ravi is the CEO of a Company. He is very powerful public speaker also."
```

识别出的词性结果如下：

```
[('Ravi', 'NNP'), ('is', 'VBZ'), ('the', 'DT'), ('CEO', 'NNP'), ('of',
'IN'), ('a', 'DT'), ('Company', 'NNP'), ('.', '.')]
[('He', 'PRP'), ('is', 'VBZ'), ('very', 'RB'), ('powerful', 'JJ'),
('public', 'JJ'), ('speaker', 'NN'), ('also', 'RB'), ('.', '.')]
```

随后，我们能够利用这些信息抽取出名词短语。

基于上述的词性输出，得出以下观察结果：

- 组块是一个或多个连续的 NNP
- 组块是一个 DT 后紧跟 NNP
- 组块是一个或多个 JJ 后紧跟 NN

通过上述三个简单的观察结论，我们可以利用词性来编写正则表达式，这种 BNF 格式看起来像标签短语：

```
NP -> <PRP>
NP -> <DT>*<NNP>
NP -> <JJ>*<NN>
NP -> <NNP>+
```

我们想要从输入文本中抽取的是下面这些组块：

- `Ravi`
- `the CEO`
- `a company`
- `powerful public speaker`

下面来编写一个简单的 Python 程序来完成任务。

6.3.1 准备工作

首先，你已经安装了 Python 和 NLTK 库。预先了解正则表达式也是非常有帮助的。

6.3.2 如何实现

1. 打开 Atom 编辑器（或者你常用的程序编辑器）。
2. 创建名为 SimpleChunker.py 的新文件。
3. 输入下列源代码：

```
SimpleChunker.py
1  import nltk
2
3  text = "Ravi is the CEO of a Company. He is very powerful public speaker also."
4
5  grammar = '\n'.join([
6      'NP: {<DT>*<NNP>}',
7      'NP: {<JJ>*<NN>}',
8      'NP: {<NNP>+}',
9  ])
10
11 sentences = nltk.sent_tokenize(text)
12
13 for sentence in sentences:
14     words = nltk.word_tokenize(sentence)
15     tags = nltk.pos_tag(words)
16     chunkparser = nltk.RegexpParser(grammar)
17     result = chunkparser.parse(tags)
18     print(result)
19
```

4. 保存文件。
5. 使用 Python 编译器运行程序。
6. 你将看到如下输出：

```
ch6 — -bash — 98×26
nltk $ python SimpleChunker.py
(S
  (NP Ravi/NNP)
  is/VBZ
  (NP the/DT CEO/NNP)
  of/IN
  (NP a/DT Company/NNP)
  ./.)
(S
  He/PRP
  is/VBZ
  very/RB
  (NP powerful/JJ public/JJ speaker/NN)
  also/RB
  ./.)
nltk $
```

6.3.3　工作原理

现在，我们了解一下程序的工作原理。将 NLTK 库导入到当前程序中：

```
import nltk
```

将我们想要处理的句子定义为 text 变量：

```
text = "Ravi is the CEO of a Company. He is very powerful public speaker
also."
```

以下这条代码利用词性编写正则表达式，因此也称为标签模式。这些标签模式并非是随机创建的，而是根据上述标签短语例子生成的：

```
grammar = '\n'.join([
  'NP: {<DT>*<NNP>}',
  'NP: {<JJ>*<NN>}',
  'NP: {<NNP>+}',
])
```

下面理解一下这些标签模式：

- NP 由一个或多个 <DT> 后紧跟一个 <NNP> 组成
- NP 由一个或多个 <JJ> 后紧跟一个 <NN> 组成
- NP 由多个 <NNP> 组成

随着处理的文本越来越多，我们就越能发现更多类似的规则。这些规则是语言处理的独特之处。因此，为了在信息抽取方面取得更多进展，我们需要不断地实践：

```
sentences = nltk.sent_tokenize(text)
```

首先，我们利用 nltk.sent_tokenize() 函数将输入文本分割成句子。

```
for sentence in sentences:
```

这条代码遍历句子列表中的所有句子，并将每个句子分配给 sentence 变量。

```
words = nltk.word_tokenize(sentence)
```

这条代码利用 nltk.word_tokenize() 函数将句子分割成单词，并将结果赋值给 words 变量。

```
tags = nltk.pos_tag(words)
```

这条代码对单词变量（已有的单词列表）进行词性识别，并将结果放入 tags 变量中（每个单词都被正确地标记了词性）。

```
chunkparser = nltk.RegexpParser(grammar)
```

这条代码表示在之前创建的 grammar 对象上调用 nltk.RegexpParser，并将其保存到变量 chunkparser 中。

```
result = chunkparser.parse(tags)
```

利用 parse 对象解析标签，并将结果存储在 result 变量中。

```
print(result)
```

这里，我们利用 print() 函数将识别的组块在屏幕上显示出来。输出是一种树形结构，包含单词及其相应的词性。

6.4　训练分块器

本节将介绍训练过程，学习如何训练我们自己的分块器并对其进行评估。在进行训练之前，我们需要先理解所处理的数据类型。只要我们能合理地理解数据，就一定能根据我们的需求来抽取一系列信息，并对其进行训练。训练数据的一种方式是利用 IOB 标签对给定文本中的组块进行标注。

通常来说，我们首先在句子中发现不同的单词，然后识别这些单词对应的词性。接下来，当对文本进行分块时，我们需要根据这些单词在文本中出现的位置进一步标记它们。

例如下面这句话：

```
"Bill Gates announces Satya Nadella as new CEO of Microsoft"
```

一旦完成词性标注和分块，我们将看到如下输出结果：

```
Bill NNP B-PERSON
Gates NNP I-PERSON
announces NNS O
Satya NNP B-PERSON
Nadella NNP I-PERSON
as IN O
new JJ O
```

```
CEO NNP B-ROLE
of IN O
Microsoft NNP B-COMPANY
```

这就是所谓的 IOB 格式，每行由以空格键分开的三部分组成。

列	描述
IOB 的第一列	输入句子中的单词
IOB 的第二列	单词对应的词性
IOB 的第三列	识别组块的 I（组块内部词）、O（组块外部词）、B（组块的开始词），以及可以代表单词种类的后缀

让我们看看下图的 IBO 格式：

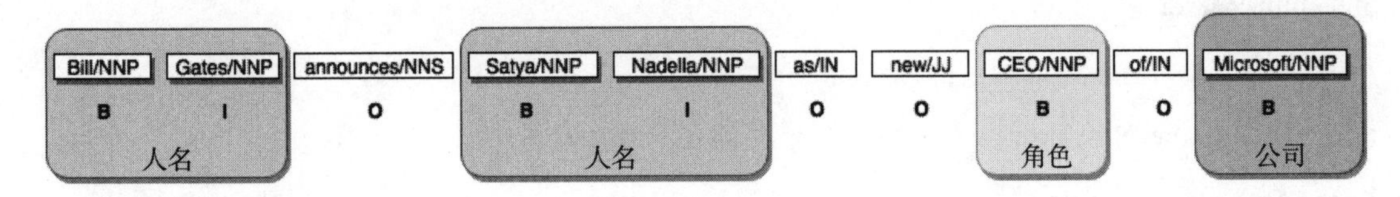

当有了 IOB 格式的训练数据，我们就能将它应用于分块器的训练，并可识别其他数据集上的组块。如果我们没有训练语料或者需要从文本中识别新的类型，训练过程将非常困难。

本节我们尝试利用 regexparser 编写一个简单的分块器并观察输出的结果类型。

6.4.1　准备工作

安装 Python 工具包以及 NLTK 库。

6.4.2　如何实现

1. 打开 Atom 编辑器（或者你常用的程序编辑器）。
2. 创建名为 TraningChunker.py 的新文件。
3. 输入下列源代码：

```
TrainingChunker.py
1  import nltk
2  from nltk.corpus import conll2000
3  from nltk.corpus import treebank_chunk
4
5  def mySimpleChunker():
6      grammar = 'NP: {<NNP>+}'
7      return nltk.RegexpParser(grammar)
8
```

```
def test_nothing(data):
    cp = nltk.RegexpParser("")
    print(cp.evaluate(data))

def test_mysimplechunker(data):
    schunker = mySimpleChunker()
    print(schunker.evaluate(data))


datasets = [
    conll2000.chunked_sents('test.txt', chunk_types=['NP']),
    treebank_chunk.chunked_sents()
]

for dataset in datasets:
    test_nothing(dataset[:50])
    test_mysimplechunker(dataset[:50])

```

4. 保存文件。
5. 使用 Python 编译器运行程序。
6. 你将看到如下输出：

```
nltk $ python TrainingChunker.py 2>/dev/null
ChunkParse score:
    IOB Accuracy:  38.6%%
    Precision:      0.0%%
    Recall:         0.0%%
    F-Measure:      0.0%%
ChunkParse score:
    IOB Accuracy:  48.2%%
    Precision:     71.1%%
    Recall:        17.2%%
    F-Measure:     27.7%%
ChunkParse score:
    IOB Accuracy:  45.0%%
    Precision:      0.0%%
    Recall:         0.0%%
    F-Measure:      0.0%%
ChunkParse score:
    IOB Accuracy:  50.7%%
    Precision:     51.9%%
    Recall:         8.8%%
    F-Measure:     15.1%%
nltk $
```

6.4.3　工作原理

这个代码将 NLTK 模块导入到当前程序中：

```
import nltk
```

这个代码将 conll2000 语料库导入到当前程序中：

```
from nltk.corpus import conll2000
```

这个代码将 treebank 语料库导入到当前程序中：

```
from nltk.corpus import treebank_chunk
```

定义一个新函数 mySimpleChunker()，同时，也定义一个简单的标签格式，它能够抽取所有词性为 NNP（专有名词）的单词。我们的分块器利用这个文法格式抽取命名实体：

```
def mySimpleChunker():
  grammar = 'NP: {<NNP>+}'
  return nltk.RegexpParser(grammar)
```

以下是一个简单的分块器，它从给定的文本中不抽取任何东西，只用于检测算法能否正常运行：

```
def test_nothing(data):
  cp = nltk.RegexpParser("")
  print(cp.evaluate(data))
```

以下代码在测试集上使用 mySimpleChunker() 函数，并根据已经标记的标准答案对分块结果的准确率进行评估：

```
def test_mysimplechunker(data):
  schunker = mySimpleChunker()
  print(schunker.evaluate(data))
```

创建两个数据集列表，一个来自 conll2000 的语料，另一个来自 treebank 的语料：

```
datasets = [
  conll2000.chunked_sents('test.txt', chunk_types=['NP']),
  treebank_chunk.chunked_sents()
]
```

遍历这两个数据集，每次在前 50 个 IOB 标注语句上调用 test_nothing() 函数和 test_mysimplechunker() 函数，并计算分块器的准确率：

```
for dataset in datasets:
  test_nothing(dataset[:50])
  test_mysimplechunker(dataset[:50])
```

6.5 递归下降句法分析

递归下降分析器是一类句法分析器，它能从左到右读取输入文本，也能生成从上到下形式的分析树，并通过先序遍历形式访问每一个结点。由于通常采用上下文无关文法来表示文本，所以分析过程是递归的。这种分析技术可用于构造分析程序语言代码的编译器。

在本节中，我们将探讨如何使用来自 NLTK 库的 RD 分析器。

6.5.1　准备工作

安装 Python 工具包以及 NLTK 库。

6.5.2　如何实现

1. 打开 Atom 编辑器（或者你常用的程序编辑器）。
2. 创建名为 ParsingRD.py 的新文件。
3. 输入下列源代码：

```python
import nltk

def RDParserExample(grammar, textlist):
    parser = nltk.parse.RecursiveDescentParser(grammar)
    for text in textlist:
        sentence = nltk.word_tokenize(text)
        for tree in parser.parse(sentence):
            print(tree)
            tree.draw()

grammar = nltk.CFG.fromstring("""
S -> NP VP
NP -> NNP VBZ
VP -> IN NNP | DT NN IN NNP
NNP -> 'Tajmahal' | 'Agra' | 'Bangalore' | 'Karnataka'
VBZ -> 'is'
IN -> 'in' | 'of'
DT -> 'the'
NN -> 'capital'
""")

text = [
    "Tajmahal is in Agra",
    "Bangalore is the capital of Karnataka",
]

RDParserExample(grammar, text)
```

4. 保存文件。
5. 使用 Python 编译器运行程序。
6. 你将看到如下输出：

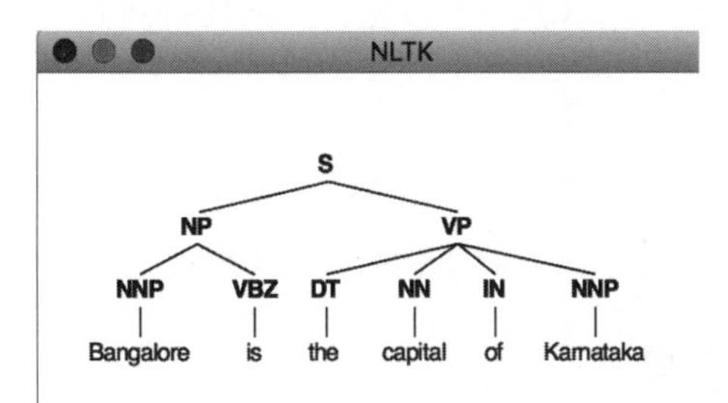

该图是 RD 分析器解析第二条输入句子的输出结果。

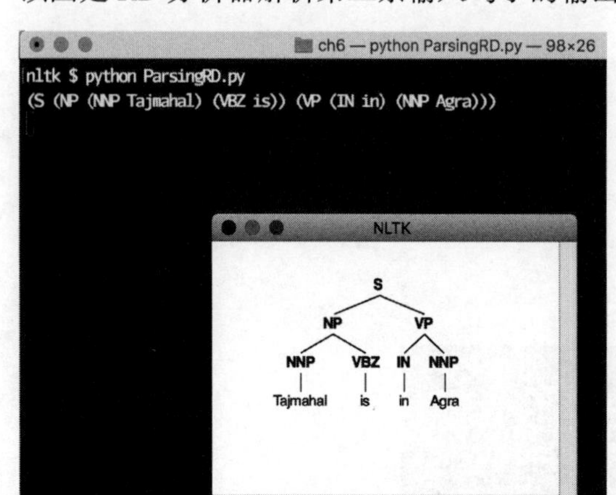

6.5.3　工作原理

首先，让我们来看程序的工作原理。以下代码用于导入 NLTK 库：

```
import nltk
```

在下面这条代码中，我们定义了一个新的函数 RDParserExample，它以 grammar 和 textlist 为参数：

```
def RDParserExample(grammar, textlist):
```

通过从 nltk.parse 库中调用 RecursiveDescentParser，我们建立了一个新的分析器。将 grammar 传递给这个类来进行初始化：

```
parser = nltk.parse.RecursiveDescentParser(grammar)
```

在下面这几条代码中，我们遍历 textlist 变量中的文本列表。利用函数 nltk.word_tokenize() 对每个文本进行分词，然后将分词结果传递给 parser.parse() 函数。当句法分析完成时，在屏幕上展示结果和分析树：

```
for text in textlist:
  sentence = nltk.word_tokenize(text)
  for tree in parser.parse(sentence):
    print(tree)
    tree.draw()
```

利用 grammar 创建一个新的 CFG 对象：

```
grammar = nltk.CFG.fromstring("""
S -> NP VP
NP -> NNP VBZ
VP -> IN NNP | DT NN IN NNP
NNP -> 'Tajmahal' | 'Agra' | 'Bangalore' | 'Karnataka'
VBZ -> 'is'
```

```
IN -> 'in' | 'of'
DT -> 'the'
NN -> 'capital'
""")
```

以下两条例句是用于理解这个句法分析器：

```
text = [
  "Tajmahal is in Agra",
  "Bangalore is the capital of Karnataka",
]
```

调用 RDParserExample，输入是 grammar 对象和例句列表：

```
RDParserExample(grammar, text)
```

6.6 shift-reduce 句法分析

在本节中，我们将学习如何使用和理解 shift-reduce 句法分析。

shift-reduce 句法分析器是一种特殊类型的句法分析器，它能以从左到右的单线程和自上而下的多线程方式分析输入文本。

对于输入文本中的每一个字母 / 单词，分析过程如下：

- 从输入文本中读取第一个单词，并将其压入堆栈中（移进操作）；
- 读取堆栈中的完整分析树，并通过从右到左读取产生式规则来查找哪条产生式规则能够进行归约（归约操作）；
- 重复这个过程并遍历所有的产生式规则，最后没有可用的规则，即分析失败；
- 重复这个过程直到所有的输入单词都压入堆栈并被归约，我们就认为分析成功。

在接下来的实例中，我们会看到仅有一个输入文本分析成功。

6.6.1 准备工作

安装 Python 工具包以及 NLTK 库，并且需要了解如何编写文法。

6.6.2 如何实现

1. 打开 Atom 编辑器（或者你常用的程序编辑器）。
2. 创建名为 ParsingSR.py 的新文件。
3. 输入下列源代码：

```
ParsingSR.py
import nltk

def SRParserExample(grammar, textlist):
    parser = nltk.parse.ShiftReduceParser(grammar)
    for text in textlist:
```

```
        sentence = nltk.word_tokenize(text)
        for tree in parser.parse(sentence):
            print(tree)
            tree.draw()

text = [
    "Tajmahal is in Agra",
    "Bangalore is the capital of Karnataka",
]

grammar = nltk.CFG.fromstring("""
S -> NP VP
NP -> NNP VBZ
VP -> IN NNP | DT NN IN NNP
NNP -> 'Tajmahal' | 'Agra' | 'Bangalore' | 'Karnataka'
VBZ -> 'is'
IN -> 'in' | 'of'
DT -> 'the'
NN -> 'capital'
""")

SRParserExample(grammar, text)
```

4. 保存文件。
5. 使用 Python 编译器运行程序。
6. 你将看到如下输出：

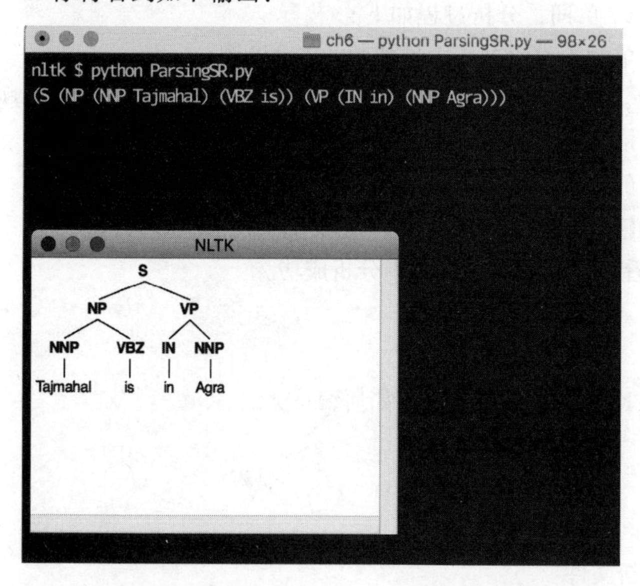

6.6.3 工作原理

首先，让我们了解一下程序的工作原理。以下代码帮助我们导入 NLTK 库：

```
import nltk
```

在下面这条代码中，我们定义了一个新的函数 SRParserExample()，它以 grammar 和

textlist 为参数：

```
def SRParserExample(grammar, textlist):
```

通过从 nltk.parse 库中调用 ShiftReduceParser，我们建立了一个新的分析器。将 grammar 传递给这个类来进行初始化：

```
parser = nltk.parse.ShiftReduceParser(grammar)
```

通过以下这几条代码，我们遍历 textlist 变量中的文本列表。利用函数 nltk.word_tokenize() 对每个文本进行分词，然后将分词结果传递给 parser.parse() 函数。当句法分析完成时，在屏幕上展示结果和分析树。

```
for text in textlist:
  sentence = nltk.word_tokenize(text)
  for tree in parser.parse(sentence):
    print(tree)
    tree.draw()
```

利用两条例句来理解 shift-reduce 分析器：

```
text = [
  "Tajmahal is in Agra",
  "Bangalore is the capital of Karnataka",
]
```

利用 grammar 创建一个新的 CFG 对象：

```
grammar = nltk.CFG.fromstring("""
S -> NP VP
NP -> NNP VBZ
VP -> IN NNP | DT NN IN NNP
NNP -> 'Tajmahal' | 'Agra' | 'Bangalore' | 'Karnataka'
VBZ -> 'is'
IN -> 'in' | 'of'
DT -> 'the'
NN -> 'capital'
""")
```

调用 SRParserExample，输入是 grammar 对象和例句列表：

```
SRParserExample(grammar, text)
```

6.7 依存句法分析和主观依存分析

在本节中，我们将学习如何进行依存句法分析以及如何使用主观依存分析器。

依存句法基于这样一个假设，即构成句子的单词之间有时存在直接关系，本节中的实例将清楚地说明这一点。

6.7.1 准备工作

安装 Python 工具包以及 NLTK 库。

6.7.2 如何实现

1. 打开 Atom 编辑器（或者你常用的程序编辑器）。
2. 创建名为 ParsingDG.py 的新文件。
3. 输入以下源代码：

```
ParsingDG.py
1  import nltk
2
3  grammar = nltk,grammar.DependencyGrammar.fromstring("""
4  'savings' -> 'small'
5  'yield' -> 'savings'
6  'gains' -> 'large'
7  'yield' -> 'gains'
8  """)
9
10 sentence = 'small savings yield large gains'
11 dp = nltk.parse.ProjectiveDependencyParser(grammar)
12 for t in sorted(dp.parse(sentence.split())):
13     print(t)
14     t.draw()
15
```

4. 保存文件。
5. 使用 Python 编译器运行程序。
6. 你将看到如下输出：

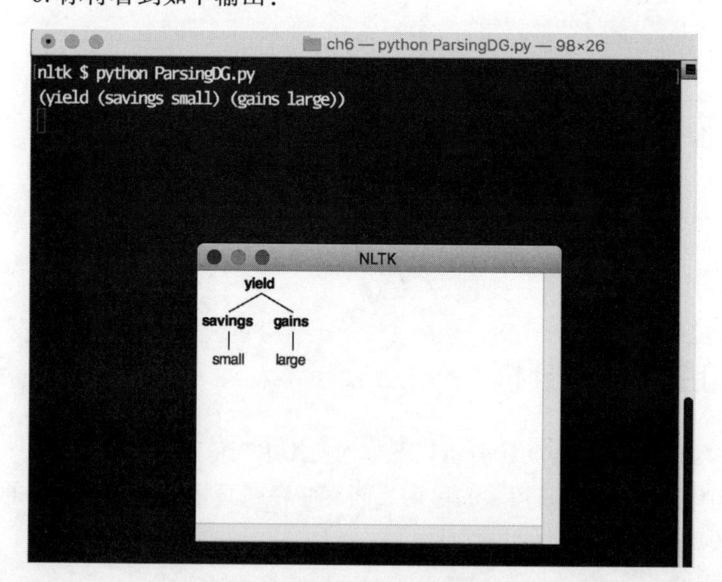

6.7.3 工作原理

首先，我们来了解一下程序的工作原理。以下代码将导入 NLTK 库：

```
import nltk
```

以下代码利用 nltk.grammar.DependencyGrammar 类创建了一个 grammar 对象。我们将下列产生式加入到 grammar 对象中：

```
grammar = nltk.grammar.DependencyGrammar.fromstring("""
'savings' -> 'small'
'yield' -> 'savings'
'gains' -> 'large'
'yield' -> 'gains'
""")
```

进一步了解这些产生式：

- small 和 savings 相关
- savings 和 yield 相关
- large 和 gains 相关
- gains 和 yield 相关

以下是需要进行句法分析的例句，它存储在一个名为 sentence 的变量中：

```
sentence = 'small savings yield large gains'
```

下面这条代码使用我们刚才定义的 grammar 对象创建了一个新的 nltk.parse.ProjectiveDependencyParser 对象：

```
dp = nltk.parse.ProjectiveDependencyParser(grammar)
```

下面这个循环完成了很多任务：

```
for t in sorted(dp.parse(sentence.split())):
  print(t)
  t.draw()
```

该循环的作用：

- 切分句子
- 分词后的所有单词赋值给 dp 对象作为输入
- 利用内置函数 sorted() 对句法分析结果进行排序
- 遍历所有的树路径，将结果展示在屏幕上，并且呈现出完整的树形结构

6.8 线图句法分析

线图句法分析器是一种特殊的句法分析器，适用于语法模糊的自然语言。线图句法分析器使用动态规划来产生预期的分析结果。

动态规划的优势在于它能将给定问题分解成子问题，并将其结果存储在一个共享位置，当任何其他地方发生类似子问题时，这个算法可以调用已存储的结果。这大大降低了相同问题的重复计算量。

在本节中，我们将学习由 NLTK 库提供的线图句法分析功能。

6.8.1 准备工作

安装 Python 工具包以及 NLTK 库，并且对文法有很好的理解。

6.8.2 如何实现

1. 打开 Atom 编辑器（或者你常用的程序编辑器）。
2. 创建名为 ParsingChart.py 的新文件。
3. 输入以下源代码：

```
ParsingChart.py
1  from nltk.grammar import CFG
2  from nltk.parse.chart import ChartParser, BU_LC_STRATEGY
3
4  grammar = CFG.fromstring("""
5  S -> T1 T4
6  T1 -> NNP VBZ
7  T2 -> DT NN
8  T3 -> IN NNP
9  T4 -> T3 | T2 T3
10 NNP -> 'Tajmahal' | 'Agra' | 'Bangalore' | 'Karnataka'
11 VBZ -> 'is'
12 IN -> 'in' | 'of'
13 DT -> 'the'
14 NN -> 'capital'
15 """)
16
17 cp = ChartParser(grammar, BU_LC_STRATEGY, trace=True)
18
19 sentence = "Bangalore is the capital of Karnataka"
20 tokens = sentence.split()
21 chart = cp.chart_parse(tokens)
22 parses = list(chart.parses(grammar.start()))
23 print("Total Edges :", len(chart.edges()))
24 for tree in parses: print(tree)
25 tree.draw()
26
```

4. 保存文件。
5. 使用 Python 编译器运行程序。
6. 你将看到如下输出：

```
nltk $ python ParsingChart.py
|.Bangal.  is  . the  .capita.  of  .Karnat.|
|[-------]      .       .       .       .       .| [0:1] 'Bangalore'
|.      [-------]       .       .       .       .| [1:2] 'is'
|.      .       [-------]       .       .       .| [2:3] 'the'
|.      .       .       [-------]       .       .| [3:4] 'capital'
|.      .       .       .       [-------]       .| [4:5] 'of'
|.      .       .       .       .       [-------]| [5:6] 'Karnataka'
|[-------]      .       .       .       .       .| [0:1] NNP -> 'Bangalore' *
|[------->      .       .       .       .       .| [0:1] T1 -> NNP * VBZ
|.      [-------]       .       .       .       .| [1:2] VBZ -> 'is' *
|[---------------]      .       .       .       .| [0:2] T1 -> NNP VBZ *
|[--------------->      .       .       .       .| [0:2] S  -> T1 * T4
|.      .       [-------]       .       .       .| [2:3] DT -> 'the' *
|.      .       [------->       .       .       .| [2:3] T2 -> DT * NN
|.      .       .       [-------]       .       .| [3:4] NN -> 'capital' *
|.      .       [---------------]       .       .| [2:4] T2 -> DT NN *
|.      .       [--------------->       .       .| [2:4] T4 -> T2 * T3
|.      .       .       .       [-------]       .| [4:5] IN -> 'of' *
|.      .       .       .       [------->       .| [4:5] T3 -> IN * NNP
|.      .       .       .       .       [-------]| [5:6] NNP -> 'Karnataka' *
|.      .       .       .       .       [------->| [5:6] T1 -> NNP * VBZ
|.      .       .       .       [---------------]| [4:6] T3 -> IN NNP *
|.      .       .       .       [---------------]| [4:6] T4 -> T3 *
|.      .       [-------------------------------]| [2:6] T4 -> T2 T3 *
|[===============================================]| [0:6] S  -> T1 T4 *
Total Edges : 24
(S
  (T1 (NNP Bangalore) (VBZ is))
  (T4 (T2 (DT the) (NN capital)) (T3 (IN of) (NNP Karnataka))))
```

6.8.3 工作原理

首先，让我们了解一下程序的工作原理。以下代码将 CFG 模块导入到程序中：

```
from nltk.grammar import CFG
```

以下代码将 ChartParser 和 BU_LC_STRATEGY 特征导入到程序中：

```
from nltk.parse.chart import ChartParser, BU_LC_STRATEGY
```

构造一个范例文法用于后续分析使用。所有的产生式都以 BNF 格式表示：

```
grammar = CFG.fromstring("""
S -> T1 T4
T1 -> NNP VBZ
T2 -> DT NN
T3 -> IN NNP
T4 -> T3 | T2 T3
NNP -> 'Tajmahal' | 'Agra' | 'Bangalore' | 'Karnataka'
VBZ -> 'is'
IN -> 'in' | 'of'
DT -> 'the'
NN -> 'capital'
""")
```

这个文法包含：

- 开始符号 S，可产生 T1 T4
- 非终结符号 T1、T2、T3 和 T4，它们分别进一步产生 NNP VBZ、DT NN、IN NNP、T2 或 T2 T3
- 终结符号，也就是英语词典中的单词

以下代码构造了一个新的线图句法分析对象，输入包括 grammar 对象、BU_LC_STRATEGY 参数，并将 trace 设置为 true，这样我们可以通过屏幕观察分析过程：

```
cp = ChartParser(grammar, BU_LC_STRATEGY, trace=True)
```

我们将在程序中处理以下字符串范例，它存储在一个名为 sentence 的变量中：

```
sentence = "Bangalore is the capital of Karnataka"
```

以下代码对范例语句创建了一个单词列表：

```
tokens = sentence.split()
```

以下代码开始句法分析，以单词列表作为输入，分析结果赋给 chart 对象：

```
chart = cp.chart_parse(tokens)
```

将 chart 对象获取到的所有分析树输入到 parses 变量中：

```
parses = list(chart.parses(grammar.start()))
```

以下代码打印出当前 chart 对象中所有边的数量：

```
print("Total Edges :", len(chart.edges()))
```

以下代码在屏幕上打印出所有的分析树：

```
for tree in parses: print(tree)
```

以下代码在图形用户界面（GUI）上显示线图分析树的视图：

```
tree.draw()
```

CHAPTER 7

第7章 信息抽取和文本分类

7.1 引言

信息检索是一个充满挑战的广阔领域。在前几章，我们已经了解了正则表达式、文法、词性标注和分块，接下来的步骤自然是从给定文本中抽取感兴趣的实体。更准确地说，当我们在处理大规模数据时，我们非常关注其中是否包含著名的人名、地名和产品名等。这些在自然语言处理中被称为**命名实体**，接下来我们将通过例子更多地了解实体。此外，我们也将利用输入文本中的线索对大量文本进行分类，并将介绍大量的例子。

了解命名实体

迄今为止，我们已经了解了如何对文本进行句法分析、词性标注以及分块。接下来，我们需要关注的是如何识别**专有名词**，也可以称为命名实体。

命名实体帮助我们了解更多关于给定文本提及的内容，以便我们可以进一步对数据进行分类。由于命名实体包含多个单词，因此，文本中的命名实体识别有时比较困难。

我们通过以下例子来理解什么是命名实体：

句子	命名实体
Hampi is on the South Bank of Tungabhabra river	Hampi，Tungabhadra River
Paris is famous for Fashion	Paris
Burj Khalifa is one of the SKyscrapers in Dubai	Burj Khalifa，Dubai
Jeff Weiner is the CEO of LinkedIn	Jeff Weiner，LinkedIn

我们仔细观察，并试图理解：

1. 虽然 South Bank（南岸）指示了一个方向，但由于它无法唯一地指向一个实物，因此它不属于命名实体。

2. Fashion（时尚）是一个名词，但不能完全等同于一个命名实体。

3.Skycraper（摩天大楼）是一个名词，但是在哪都可能存在摩天大楼。

4.CEO（首席执行官）是一个职务，很多人可能担任这个头衔。因此，它同样也不是一个命名实体。

为了进一步理解，我们从分类的角度来看这些实体：

类别	命名实体范例
TIMEZONE	Asia/Kolkata，IST，UTC
LOCATION	Mumbai，Kolkata，Egypt
RIVERS	Ganga，Yamuna，Nile
COSMETICS	Maybelline Deep Coral Lipstick，LOreal Excellence Creme Hair Color
CURRENCY	100 bitcoins，1.25INR
DATE	17-Aug-2017，19-Feb-2016
TIME	10:10AM
PERSON	Satya Nadella，Jeff Weiner，Bill Gates

7.2 使用内置的命名实体识别工具

Python NLTK 提供了**命名实体识别工具**。为了利用这个功能，我们首先需要回顾一下目前为止我们已经完成了哪些工作：

1. 将一个大文档分割成句子。
2. 将句子分割成词。
3. 对句子进行词性标注。
4. 从句子中提取包含连续词（非重叠）的组块。
5. 给这些组块包含的词标注 IOB 标签。

接下来的第 6 步将是扩展这些算法来识别命名实体。因此，我们本章使用的数据是经过以上 5 个步骤预处理后的。

我们将利用 treebank 数据来了解命名实体识别过程，这个数据已经被预先标注为 IOB 格式。目前为止，除了训练过程我们还没使用任何算法。

为了更好地理解训练过程的重要性，我们举一个例子。比如，考古部门需要推断出基于坎纳达语的社交网站转发或提及了哪些印度名胜古迹。

假设他们已经获得了万亿字节的数据甚至是千万亿字节的数据，他们如何寻找到这些名字？在这些原始输入上，我们需要采用一个样例数据集来进行训练，并通过这些训练数据集来抽取坎纳达命名实体。

7.2.1 准备工作

安装 Python 工具包和 NLTK 库。

7.2.2 如何实现

1. 打开 Atom 编辑器（或者你常用的程序编辑器）。

2. 创建一个新文件，命名为 NER.py。
3. 输入以下源代码：

```
NER.py
import nltk

def sampleNE():
    sent = nltk.corpus.treebank.tagged_sents()[0]
    print(nltk.ne_chunk(sent))

def sampleNE2():
    sent = nltk.corpus.treebank.tagged_sents()[0]
    print(nltk.ne_chunk(sent, binary=True))

if __name__ == '__main__':
    sampleNE()
    sampleNE2()
```

4. 保存文件。
5. 使用 Python 编译器运行程序。
6. 你将看到如下输出：

```
ch7 — -bash — 99×39
nltk $ python NER.py
(S
  (PERSON Pierre/NNP)
  (ORGANIZATION Vinken/NNP)
  ,/,
  61/CD
  years/NNS
  old/JJ
  ,/,
  will/MD
  join/VB
  the/DT
  board/NN
  as/IN
  a/DT
  nonexecutive/JJ
  director/NN
  Nov./NNP
  29/CD
  ./.)
(S
  (NE Pierre/NNP Vinken/NNP)
  ,/,
  61/CD
  years/NNS
  old/JJ
  ,/,
  will/MD
  join/VB
  the/DT
  board/NN
  as/IN
  a/DT
  nonexecutive/JJ
  director/NN
  Nov./NNP
  29/CD
  ./.)
nltk $
```

7.2.3 工作原理

这些代码看起来很简单，所有的算法都在 NLTK 库中实现。那么，我们来探究这些简单的代码是如何实现我们想要的功能。以下这条代码在程序中调用 NLTK 库：

```
import nltk
```

以下三条代码定义了一个新的函数，命名为 sampleNE()。从 treebank 语料库中导入第一个标注好的句子，然后把它传递给 nltk.ne_chunk() 函数以识别命名实体。该程序的输出包含了所有的命名实体以及它们对应的实体类别：

```
def sampleNE():
  sent = nltk.corpus.treebank.tagged_sents()[0]
  print(nltk.ne_chunk(sent))
```

以下三条代码定义了一个新的函数，命名为 sampleNE2()。导入 treebank 语料库中的第一条标注好的句子，并传递给 nltk.ne_chunk() 函数以识别命名实体。该程序的输出包含了所有识别出的无类别的命名实体。这种方式适用于利用训练集无法准确性地对命名实体进行分类的情况，类别包括人名、组织机构名和地名等。

```
def sampleNE2():
  sent = nltk.corpus.treebank.tagged_sents()[0]
  print(nltk.ne_chunk(sent, binary=True))
```

以下三条代码将调用我们之前已经定义好的两个函数，并在屏幕上打印出结果。

```
if __name__ == '__main__':
  sampleNE()
  sampleNE2()
```

7.3 创建字典、逆序字典和使用字典

作为一种通用的编程语言，Python 支持多种内置的数据结构。毫无疑问，其中最强大的数据结构之一就是字典。在学习什么是字典之前，我们先试着理解这些数据结构应用于哪些方面。简而言之，数据结构有助于程序员存储、检索和遍历存储在这些结构中的数据。每种数据结构都有自己的功能和优势，这些是程序员在任务开发中选择数据结构前需要熟悉的。

首先，我们用一个简单的例子来介绍一下字典这个数据结构的基本使用方法：

```
All the flights got delayed due to bad weather
```

我们可以对上述例句进行词性识别。也许有人想知道，句子中 flight 的词性是什么？我们应该有一个高效的方法来查找这个词。这时字典就发挥了作用。我们可以把字典看作是对所关注的数据一种**一对一**（one-to-one）映射。这种一对一的关系从最高级别上抽象了我们所关注的数据单元。如果你是一个精通 Python 程序的高手，你还能掌握**多对多** (many-tomany) 的关系。在本例中，我们希望使用字典能够获得：

```
flights -> Noun
Weather -> Noun
```

现在我们来回答另一个问题。如果我们需要在屏幕上输出给定句子中所有词性为名词的单词，是否能够实现？答案是肯定的，因为我们将学习如何使用 Python 的字典。

7.3.1 准备工作

为了实现本节的任务，你的计算机上需要安装 Python 和 NLTK 库。

7.3.2 如何实现

1. 打开 Atom 编辑器（或者你常用的程序编辑器）。
2. 新建一个文件，命名为 Dictionary.py。
3. 输入以下源代码：

```
Dictionary.py
1  import nltk
2
3  class LearningDictionary():
4      def __init__(self, sentence):
5          self.words = nltk.word_tokenize(sentence)
6          self.tagged = nltk.pos_tag(self.words)
7          self.buildDictionary()
8          self.buildReverseDictionary()
9
10     def buildDictionary(self):
11         self.dictionary = {}
12         for (word, pos) in self.tagged:
13             self.dictionary[word] = pos
14
15     def buildReverseDictionary(self):
16         self.rdictionary = {}
17         for key in self.dictionary.keys():
18             value = self.dictionary[key]
19             if value not in self.rdictionary:
20                 self.rdictionary[value] = [key]
21             else:
22                 self.rdictionary[value].append(key)
23
24     def isWordPresent(self, word):
25         return 'Yes' if word in self.dictionary else 'No'
26
27
28     def getPOSForWord(self, word):
29         return self.dictionary[word] if word in self.dictionary else None
30
31     def getWordsForPOS(self, pos):
32         return self.rdictionary[pos] if pos in self.rdictionary else None
33
34
```

```
sentence = "All the flights got delayed due to bad weather"
learning = LearningDictionary(sentence)
words = ["chair", "flights", "delayed", "pencil", "weather"]
pos = ["NN", "VBS", "NNS"]
for word in words:
    status = learning.isWordPresent(word)
    print("Is '{}' present in dictionary ? : '{}'".format(word, status))
    if status is True:
        print("\tPOS For '{}' is '{}'".format(word,
        learning.getPOSForWord(word)))
for pword in pos:
    print("POS '{}' has '{}' words".format(pword,
    learning.getWordsForPOS(pword)))
```

4. 保存文件。
5. 使用 Python 编译器运行程序。
6. 你将看到如下输出：

```
nltk $ python Dictionary.py
Is 'chair' present in dictionary ? : 'No'
Is 'flights' present in dictionary ? : 'Yes'
Is 'delayed' present in dictionary ? : 'Yes'
Is 'pencil' present in dictionary ? : 'No'
Is 'weather' present in dictionary ? : 'Yes'
POS 'NN' has '['weather']' words
POS 'VBS' has 'None' words
POS 'NNS' has '['flights']' words
nltk $
```

7.3.3　工作原理

通过前面的小节我们学习了字典的基本使用方法，现在我们来更深入地了解字典。首先导入 NLTK 库，代码如下：

```
import nltk
```

然后，定义一个新的类，类名为 LearningDictionary。代码如下：

```
class LearningDictionary():
```

为该类定义一个新的构造方法，并将原始句子文本作为该方法的参数。代码如下：

```
def __init__(self, sentence):
```

调用 nltk.word_tokenize() 函数将原始句子切分成单词列表，并将函数返回的单词列表保存在 words 类成员变量中：

```
self.words = nltk.word_tokenize(sentence)
```

在提取出单词列表之后，我们对这个列表的每一个单词进行词性标注，并将结果保存在类成员变量 tagged 中。代码如下：

```
self.tagged = nltk.pos_tag(self.words)
```

然后我们调用 buildDictionary() 函数，该函数在类中已经定义。代码如下：

```
self.buildDictionary()
```

再调用 buildReverseDictionary() 函数，该函数同样已经在类中定义。代码如下：

```
self.buildReverseDictionary()
```

以下代码定义了一个新的类成员方法，命名为 buildDictionary()：

```
def buildDictionary(self):
```

以下两条代码在类中初始化一个空的 dictionary 变量，然后迭代 tagged 变量中所有的 pos（词性）元素，并将每一个词性和它对应的 word（单词）加入字典，将 word 作为字典的键，pos 作为对应的键值：

```
self.dictionary = {}
for (word, pos) in self.tagged:
  self.dictionary[word] = pos
```

定义另一个类成员方法，命名为 buildReverseDictionary()。代码如下：

```
def buildReverseDictionary(self):
```

再创建一个空字典作为类成员变量，命名为 rdictionary()。代码如下：

```
self.rdictionary = {}
```

以下代码遍历字典中所有的键，遍历到每个键时都将其赋值给名为 key 的局部变量：

```
for key in self.dictionary.keys():
```

每遍历到一个键（即原始单词）时，我们都会获取到其对应的键值（词性），并将其存储在名为 value 的局部变量中。代码如下：

```
value = self.dictionary[key]
```

接下来，检查给定的键是否已经存在于逆序字典变量（即 rdictionary 变量）中。如果存在，将当前找到的单词添加到列表的末尾；如果不存在，建立一个表长为 1 的新列表，并将这个单词存入列表中。代码如下：

```
if value not in self.rdictionary:
  self.rdictionary[value] = [key]
else:
  self.rdictionary[value].append(key)
```

下列函数用于判断给定的单词在 dictionary 变量中是否存在，如果存在，则返回 Yes，否则返回 No：

```
def isWordPresent(self, word):
  return 'Yes' if word in self.dictionary else 'No'
```

再定义一个 getPOSForWord 函数，该函数的功能是通过查找 dictionary 变量返回给定单词的词性。如果找不到，则返回 None。代码如下：

```
def getPOSForWord(self, word):
  return self.dictionary[word] if word in self.dictionary else None
```

定义一个 getWordsForPOS 函数，该函数的功能是通过查找 rdictionary（即逆序字典变量），返回句子中每个单词的词性。如果查找不到词性，则返回 None 值：

```
def getWordsForPOS(self, pos):
  return self.rdictionary[pos] if pos in self.rdictionary else None
```

定义一个名为 sentence 的变量，并将我们想要分析的句子保存到这个变量中。代码如下：

```
sentence = "All the flights got delayed due to bad weather"
```

初始化 LearningDictionary() 类，这个类以变量 sentence 为参数，并被保存到名为 learning 的变量中：

```
learning = LearningDictionary(sentence)
```

创建一个 words 列表，并将需要词性标注的单词保存到列表中。我们在 words 列表中加入了一些在 sentence 变量中没有出现的单词，如下列代码所示：

```
words = ["chair", "flights", "delayed", "pencil", "weather"]
```

与此同时，我们也创建了一个 pos 列表，包含我们想要对单词进行分类的词性。代码如下：

```
pos = ["NN", "VBS", "NNS"]
```

以下代码遍历 words 列表中的每一个单词，并将每次遍历到的元素都保存到 word 变量中，再检查这个单词在字典中是否存在（检查的方法是调用 learning 对象的 isWordPresent() 函数，通过函数的返回值来判断），然后在屏幕上输出状态信息。如果可以在字典中找到这个单词，就将它对应的词性输出到屏幕：

```
for word in words:
  status = learning.isWordPresent(word)
  print("Is '{}' present in dictionary ? : '{}'".format(word, status))
```

```
  if status is True:
    print("\tPOS For '{}' is '{}'".format(word,
learning.getPOSForWord(word)))
```

接着，遍历 pos 列表中的所有元素，并将每次遍历到的元素都保存到 pword 变量中，再调用 getWordsForPOS() 函数将所有词性为 pword 的单词一一输出到屏幕上。代码如下：

```
for pword in pos:
  print("POS '{}' has '{}' words".format(pword,
learning.getWordsForPOS(pword)))
```

7.4 特征集合选择

特征是 NLTK 库最重要的组成部分之一。当需要为文本加标签时，特征可以为我们提供语言中的线索，使数据的标注更加轻松。在 Python 程序中，特征通常以字典的形式存在，字典中的键代表标签，而键值代表从输入数据中提取的特征。

这里我们以运输部的数据为例，如果给定一辆车的编号，我们能否判断出这辆车是否为卡纳塔克邦政府所有。在找不到给定数据的任何线索的情况下，我们该如何对车辆编号进行分类呢？

让我们来了解一下车辆编号与车辆类型存在什么关系，列表如下：

车辆编号	车辆类型
KA-[0-9]{2} [0-9]{2}	普通车辆编号
KA-[0-9]{2}-F	KSRTC 和 BMTC 车辆
KA-[0-9]{2}-G	政府车辆

有了这些线索（即特征），我们可以编写一个小程序，根据车辆编号实现对车辆分类的功能。

7.4.1 准备工作

为了实现本节的任务，你的计算机上需要安装 Python 和 NLTK 库。

7.4.2 如何实现

1. 打开 Atom 编辑器（或者你常用的程序编辑器）。
2. 创建一个新文件，命名为 Features.py。
3. 输入以下源代码：

```
Features.py
1  import nltk
2  import random
3
```

```
sampledata = [
    ('KA-01-F 1034 A', 'rtc'),
    ('KA-02-F 1030 B', 'rtc'),
    ('KA-03-FA 1200 C', 'rtc'),
    ('KA-01-G 0001 A', 'gov'),
    ('KA-02-G 1004 A', 'gov'),
    ('KA-03-G 0204 A', 'gov'),
    ('KA-04-G 9230 A', 'gov'),
    ('KA-27 1290', 'oth')
]
random.shuffle(sampledata)
testdata = [
    'KA-01-G 0109',
    'KA-02-F 9020 AC',
    'KA-02-FA 0801',
    'KA-01 9129'
]
def learnSimpleFeatures():
    def vehicleNumberFeature(vnumber):
        return {'vehicle_class': vnumber[6]}
    featuresets = [(vehicleNumberFeature(vn), cls) for (vn, cls) in sampledata]
    classifier = nltk.NaiveBayesClassifier.train(featuresets)
    for num in testdata:
        feature = vehicleNumberFeature(num)
        print("(simple) %s is of type %s" %(num, classifier.classify(feature)))

def learnFeatures():
    def vehicleNumberFeature(vnumber):
        return {
            'vehicle_class': vnumber[6],
            'vehicle_prev': vnumber[5]
        }
    featuresets = [(vehicleNumberFeature(vn), cls) for (vn, cls) in sampledata]
    classifier = nltk.NaiveBayesClassifier.train(featuresets)
    for num in testdata:
        feature = vehicleNumberFeature(num)
        print("(dual) %s is of type %s" %(num, classifier.classify(feature)))

learnSimpleFeatures()
learnFeatures()
```

4. 保存文件。
5. 使用 Python 编译器运行程序。
6. 你将看到如下输出：

```
ch7 — -bash — 95×24
nltk $ python Features.py
(simple) KA-01-G 0109 is of type gov
(simple) KA-02-F 9020 AC is of type rtc
(simple) KA-02-FA 0801 is of type rtc
(simple) KA-01 9129 is of type gov
(dual) KA-01-G 0109 is of type gov
(dual) KA-02-F 9020 AC is of type rtc
(dual) KA-02-FA 0801 is of type rtc
(dual) KA-01 9129 is of type oth
nltk $
```

7.4.3 工作原理

我们来理解代码是如何实现的。首先，导入 NLTK 库和 random 库，代码如下：

```
import nltk
import random
```

然后建立一个 Python 列表，表中元素为元组。元组中包含两个元素，第一个元素为车辆编号，第二个元素为编号的预定义标签。

我们定义所有的车辆编号都只能属于下面三类中的一类：rtc、gov 和 oth。代码如下：

```
sampledata = [
  ('KA-01-F 1034 A', 'rtc'),
  ('KA-02-F 1030 B', 'rtc'),
  ('KA-03-FA 1200 C', 'rtc'),
  ('KA-01-G 0001 A', 'gov'),
  ('KA-02-G 1004 A', 'gov'),
  ('KA-03-G 0204 A', 'gov'),
  ('KA-04-G 9230 A', 'gov'),
  ('KA-27 1290', 'oth')
]
```

把 sampledata 列表中的元素随机排序（这里用到了 random 类中的 shuffle() 方法），以确保算法的输出结果不会因为元素的输入顺序而出现偏差。代码如下：

```
random.shuffle(sampledata)
```

设置一个测试数据集，命名为 testdata，该数据集是一个列表，将用于后续对车辆类别的检验。代码如下：

```
testdata = [
  'KA-01-G 0109',
  'KA-02-F 9020 AC',
  'KA-02-FA 0801',
  'KA-01 9129'
]
```

定义一个新函数，命名为 learnSimpleFeatures()，代码如下：

```
def learnSimpleFeatures():
```

以下代码又定义了一个新的函数 vehicleNumberFeature()，该函数能够获取车辆编号，并返回编号字符串的第 7 个字符，返回值类型为字典类型。我们可以利用该函数获取数据的特征。代码如下：

```
def vehicleNumberFeature(vnumber):
  return {'vehicle_class': vnumber[6]}
```

建立一个新的特征元组的列表，命名为 featuresets，表中元素为元组。元组由两部分组成，第一部分为提取的特征字典，第二部分为数据标签。值得注意的是，一旦执行了这行代码，sampledata 变量中的输入车辆编号将不再可视化，请读者切记。代码如下所示：

```
featuresets = [(vehicleNumberFeature(vn), cls) for (vn, cls) in sampledata]
```

使用上述的特征列表 featuresets 中的特征字典和数据标签来训练一个朴素贝叶斯分类器（NaiveBayesClassifier）。训练完成后的分类器会保存至 classifier 对象中，后续我们会用到该对象。代码如下：

```
classifier = nltk.NaiveBayesClassifier.train(featuresets)
```

接下来我们开始进行测试工作。首先遍历测试数据集，调用 vehicleNumberFeature() 函数来获取数据的特征，并保存到名为 feature 的变量中，然后利用上面训练好的分类器进行测试。通过观察结果不难发现，我们先前编写的特征提取函数效果欠佳，它并未正确地为编号分类。代码如下：

```
for num in testdata:
  feature = vehicleNumberFeature(num)
  print("(simple) %s is of type %s" %(num, classifier.classify(feature)))
```

为此，我们不妨对该函数做一些改进，首先定义一个新函数，命名为 learnFeatures()，代码如下：

```
def learnFeatures():
```

然后，改变特征获取的方法，即重新定义 vehicleNumberFeature() 函数。调整函数的返回值，使函数由原来的返回一个键变为返回两个键：第一个键的命名不变，即 vehicle_class，其键值为车辆编号字符串中第 7 个字符；第二个键命名为 vehicle_prev，其键值为车辆编号字符串中第 6 个的字符。这样的调整对特征提取进行了优化。代码如下：

```
def vehicleNumberFeature(vnumber):
  return {
    'vehicle_class': vnumber[6],
    'vehicle_prev': vnumber[5]
  }
```

有了重新定义的 vehicleNumberFeature() 函数，再根据给定的训练数据生成 featureset 列表。和上述介绍一致，一旦执行了这行代码，原始输入车辆编号将不再可见。代码如下：

```
featuresets = [(vehicleNumberFeature(vn), cls) for (vn, cls) in sampledata]
```

利用新的 featuresets 列表重新训练朴素贝叶斯分类器，并返回训练完成的分类器模型，将来我们会用到这一模型。代码如下：

```
classifier = nltk.NaiveBayesClassifier.train(featuresets)
```

然后，同样对分类器进行测试，测试方法与第一次实验相同。首先循环遍历 testdata 变量中的元素，提取出每个数据的特征，然后用训练好的新模型分类，输出输入车辆编号的分类。代码如下：

```
for num in testdata:
  feature = vehicleNumberFeature(num)
  print("(dual) %s is of type %s" %(num, classifier.classify(feature)))
```

调用这两个函数，并在屏幕上输出结果。代码如下：

```
learnSimpleFeatures()
learnFeatures()
```

观察结果，我们不难看出，第一个函数的输出中有一个假阳性错误，即无法识别标签为 gov 的车辆。而第二个函数效果更好，这是因为第二个模型在训练时提取了更多的特征，因此准确率也得到了有效提升。

7.5 利用分类器分割句子

如果原始句子中含有问号（?）、句号（.）和感叹号（!），那么判断这些标点符号后的句子表达是否结束，对我们来说这依然是一个很大的挑战。

这是一个有待解决的经典问题。

为了解决这个问题，我们首先要设法找出特征（或线索），并利用这些特征（或线索）来设计分类器，使其能够从大文本中提取出句子。

TIP 如果原始句子后以符号“.”为标记，并且符号“.”后第一个单词的首字母为大写，则可以判断“.”前的句子表达已经结束。

利用这两项特征，我们尝试设计一个简单分类器对句子进行标记。

7.5.1 准备工作

为了完成本节的任务，你的计算机需要安装 Python 和 NLTK 库。

7.5.2 如何实现

1. 打开 Atom 编辑器（或者你常用的程序编辑器）。
2. 创建一个新文件，命名为 Segmentation.py。
3. 输入以下源代码：

```
Segmentation.py
1 import nltk
2 def featureExtractor(words, i):
3     return ({'current-word': words[i], 'next-is-upper':
      words[i+1][0].isupper()}, words[i+1][0].isupper())
4 def getFeaturesets(sentence):
5     words = nltk.word_tokenize(sentence)
6     featuresets = [featureExtractor(words, i) for i in range(1, len(words) - 1)
      if words[i] == '.']
7     return featuresets
```

```
def segmentTextAndPrintSentences(data):
    words = nltk.word_tokenize(data)
    for i in range(0, len(words) - 1):
        if words[i] == '.':
            if classifier.classify(featureExtractor(words, i)[0]) == True:
                print(".")
            else:
                print(words[i], end='')
        else:
            print("{} ".format(words[i]), end='')
    print(words[-1])
# copied the text from https://en.wikipedia.org/wiki/India
traindata = "India, officially the Republic of India (Bhārat Gaṇarājya),[e] is a country in South Asia. it is the seventh-largest country by area, the second-most populous country (with over 1.2 billion people), and the most populous democracy in the world. It is bounded by the Indian Ocean on the south, the Arabian Sea on the southwest, and the Bay of Bengal on the southeast. It shares land borders with Pakistan to the west;[f] China, Nepal, and Bhutan to the northeast; and Myanmar (Burma) and Bangladesh to the east. In the Indian Ocean, India is in the vicinity of Sri Lanka and the Maldives. India's Andaman and Nicobar Islands share a maritime border with Thailand and Indonesia."
testdata = "The Indian subcontinent was home to the urban Indus Valley Civilisation of the 3rd millennium BCE. In the following millennium, the oldest scriptures associated with Hinduism began to be composed. Social stratification, based on caste, emerged in the first millennium BCE, and Buddhism and Jainism arose. Early political consolidations took place under the Maurya and Gupta empires; the later peninsular Middle Kingdoms influenced cultures as far as southeast Asia. In the medieval era, Judaism, Zoroastrianism, Christianity, and Islam arrived, and Sikhism emerged, all adding to the region's diverse culture. Much of the north fell to the Delhi sultanate; the south was united under the Vijayanagara Empire. The economy expanded in the 17th century in the Mughal Empire. In the mid-18th century, the subcontinent came under British East India Company rule, and in the mid-19th under British crown rule. A nationalist movement emerged in the late 19th century, which later, under Mahatma Gandhi, was noted for nonviolent resistance and led to India's independence in 1947."

traindataset = getFeaturesets(traindata)
classifier = nltk.NaiveBayesClassifier.train(traindataset)
segmentTextAndPrintSentences(testdata)
```

4. 保存文件。

5. 使用 Python 编译器运行程序。

6. 你将看到如下输出：

```
ch7 — -bash — 95×24
nltk $ python Segmentation.py
The Indian subcontinent was home to the urban Indus Valley Civilisation of the 3rd millennium B
CE .
In the following millennium , the oldest scriptures associated with Hinduism began to be compos
ed .
Social stratification , based on caste , emerged in the first millennium BCE , and Buddhism and
 Jainism arose .
Early political consolidations took place under the Maurya and Gupta empires ; the later penins
ular Middle Kingdoms influenced cultures as far as southeast Asia .
In the medieval era , Judaism , Zoroastrianism , Christianity , and Islam arrived , and Sikhism
 emerged , all adding to the region 's diverse culture .
Much of the north fell to the Delhi sultanate ; the south was united under the Vijayanagara Emp
ire .
The economy expanded in the 17th century in the Mughal Empire .
In the mid-18th century , the subcontinent came under British East India Company rule , and in
the mid-19th under British crown rule .
A nationalist movement emerged in the late 19th century , which later , under Mahatma Gandhi ,
was noted for nonviolent resistance and led to India 's independence in 1947 .
nltk $
```

7.5.3 工作原理

我们来理解代码是如何实现的。首先导入了 NLTK 库。代码如下：

```
import nltk
```

然后，定义一个改进的特征提取函数，它返回一个包含特征字典的元组，并返回 True 或 False 来判断特征是否能确定一个句子边界。代码如下：

```
def featureExtractor(words, i):
    return ({'current-word': words[i], 'next-is-upper':
words[i+1][0].isupper()}, words[i+1][0].isupper())
```

我们再来定义一个函数，该函数以 sentence 变量为参数，返回包含元组的 featuresets 列表，元组中的元素是特征字典和布尔值。代码如下：

```
def getFeaturesets(sentence):
  words = nltk.word_tokenize(sentence)
  featuresets = [featureExtractor(words, i) for i in range(1, len(words) -
1) if words[i] == '.']
  return featuresets
```

接着定义一个函数，该函数以原始文本为参数，其功能是对文本分词，并将每个单词保存至列表中，然后遍历该列表。如果在遍历的过程中遇到句号，函数将调用分类器来判断这个位置是否为句末。如果分类器的返回值为 True，那么这句话就结束了。然后函数会继续遍历下一个单词。这个过程会一直持续到文本结束为止，代码如下：

```
def segmentTextAndPrintSentences(data):
  words = nltk.word_tokenize(data)
  for i in range(0, len(words) - 1):
    if words[i] == '.':
      if classifier.classify(featureExtractor(words, i)[0]) == True:
        print(".")
      else:
        print(words[i], end='')
    else:
      print("{} ".format(words[i]), end='')
    print(words[-1])
```

下面给出训练数据和测试数据，训练数据用于模型的训练，测试数据用来验证分类器的准确率。代码如下：

```
# copied the text from https://en.wikipedia.org/wiki/India
traindata = "India, officially the Republic of India (Bhārat Gaṇarājya),[e]
is a country in South Asia. it is the seventh-largest country by area, the
second-most populous country (with over 1.2 billion people), and the most
populous democracy in the world. It is bounded by the Indian Ocean on the
south, the Arabian Sea on the southwest, and the Bay of Bengal on the
southeast. It shares land borders with Pakistan to the west;[f] China,
Nepal, and Bhutan to the northeast; and Myanmar (Burma) and Bangladesh to
the east. In the Indian Ocean, India is in the vicinity of Sri Lanka and
the Maldives. India's Andaman and Nicobar Islands share a maritime border
with Thailand and Indonesia."
```

```
testdata = "The Indian subcontinent was home to the urban Indus Valley
Civilisation of the 3rd millennium BCE. In the following millennium, the
oldest scriptures associated with Hinduism began to be composed. Social
stratification, based on caste, emerged in the first millennium BCE, and
Buddhism and Jainism arose. Early political consolidations took place under
the Maurya and Gupta empires; the later peninsular Middle Kingdoms
influenced cultures as far as southeast Asia. In the medieval era, Judaism,
Zoroastrianism, Christianity, and Islam arrived, and Sikhism emerged, all
adding to the region's diverse culture. Much of the north fell to the Delhi
sultanate; the south was united under the Vijayanagara Empire. The economy
expanded in the 17th century in the Mughal Empire. In the mid-18th century,
the subcontinent came under British East India Company rule, and in the
mid-19th under British crown rule. A nationalist movement emerged in the
late 19th century, which later, under Mahatma Gandhi, was noted for
nonviolent resistance and led to India's independence in 1947."
```

首先，从训练数据（traindata 变量）中提取所有的特征，并将其保存至 traindataset 变量中。代码如下：

```
traindataset = getFeaturesets(traindata)
```

然后，调用 NaiveBayesClassifier（朴素贝叶斯分类器）在 traindataset 变量上进行训练，得到 classifier（分类器）对象。代码如下：

```
classifier = nltk.NaiveBayesClassifier.train(traindataset)
```

最后，调用函数在 testdata 变量上进行测试，并将最终切分出的句子输出到屏幕。代码如下：

```
segmentTextAndPrintSentences(testdata)
```

7.6　文本分类

在本节中，我们将介绍如何设计一个可用于文本分类的分类器。本节我们以丰富站点（Rich Site Summary，RSS）源的分类为例。分类的类别标签是已知的，这一点对分类任务来说很重要。

当今是一个信息化的时代，互联网上已有大量可用的文本。依靠人工对这些文本进行分类是不现实的。而在训练数据上训练生成的分类模型可以帮助我们对这些新出现的文本进行正确的分类。

7.6.1　准备工作

为了完成本节的任务，你的计算机需要安装 Python 和 NLTK 库。

7.6.2　如何实现

1. 打开 Atom 编辑器（或者你常用的程序编辑器）。

2. 创建一个新文件，命名为 DocumentClassify.py。

3. 输入以下源代码：

```
DocumentClassify.py
import nltk
import random
import feedparser

urls = {
    'mlb': 'https://sports.yahoo.com/mlb/rss.xml',
    'nfl': 'https://sports.yahoo.com/nfl/rss.xml',
}

feedmap = {}
stopwords = nltk.corpus.stopwords.words('english')

def featureExtractor(words):
    features = {}
    for word in words:
        if word not in stopwords:
            features["word({})".format(word)] = True
    return features

sentences = []

for category in urls.keys():
    feedmap[category] = feedparser.parse(urls[category])
    print("downloading {}".format(urls[category]))
    for entry in feedmap[category]['entries']:
        data = entry['summary']
        words = data.split()
        sentences.append((category, words))

featuresets = [(featureExtractor(words), category) for category, words in
sentences]
random.shuffle(featuresets)

total = len(featuresets)
off = int(total/2)
trainset = featuresets[off:]
testset = featuresets[:off]

classifier = nltk.NaiveBayesClassifier.train(trainset)

print(nltk.classify.accuracy(classifier, testset))

classifier.show_most_informative_features(5)
for (i, entry) in enumerate(feedmap['nfl']['entries']):
    if i < 4:
        features = featureExtractor(entry['title'].split())
        category = classifier.classify(features)
        print('{} -> {}'.format(category, entry['summary']))
```

4. 保存文件。

5. 使用 Python 编译器运行程序。

6. 你将看到如下输出：

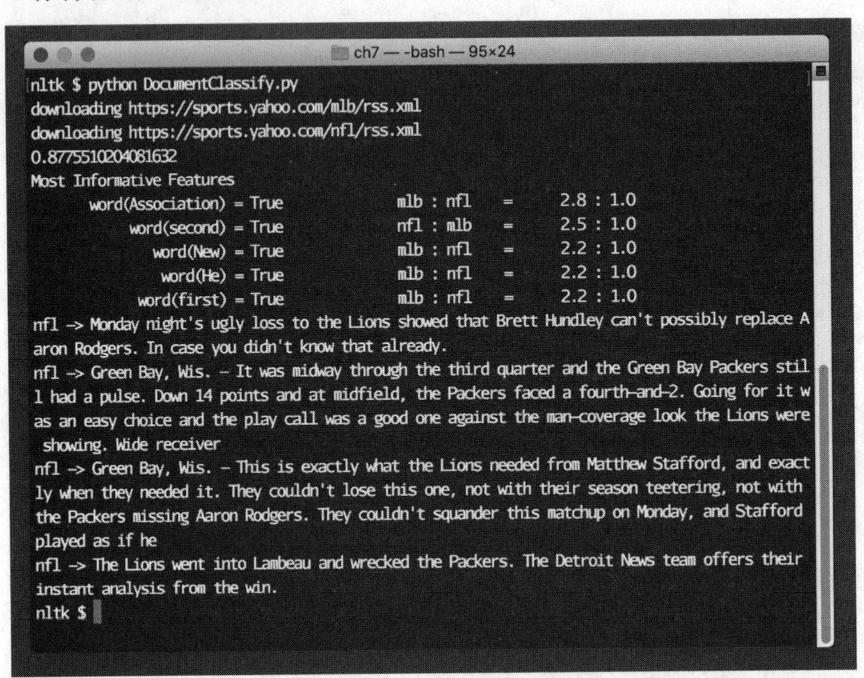

7.6.3　工作原理

我们来理解上述代码是如何实现的。首先在程序中导入 3 个库，代码如下：

```
import nltk
import random
import feedparser
```

以下代码建立了一个字典变量，变量中包含两个跟雅虎体育相关的 RSS 源，这两个源已经预先被分好类。我们为什么要选择这两个 RSS 源呢？因为在网上能够方便获取这两个源包含的内容。代码如下：

```
urls = {
  'mlb': 'https://sports.yahoo.com/mlb/rss.xml',
  'nfl': 'https://sports.yahoo.com/nfl/rss.xml',
}
```

初始化一个字典变量，命名为 feedmap，这个字典变量用于存储 RSS 源，以确保它们在程序运行结束前不会丢失。代码如下：

```
feedmap = {}
```

以下代码获取英文形式的 stopwords（停用词）列表，并将其保存在 stopwords 变量中。代码如下：

```
stopwords = nltk.corpus.stopwords.words('english')
```

定义一个 featureExtractor () 函数。该函数的输入参数为单词列表，并将它们添加到 features 字典中，使每个非停用词作为字典的键，而所有键对应的键值均为 True。函数运行结束后返回 features 字典，该字典便是给定输入 words 的特征。代码如下：

```
def featureExtractor(words):
  features = {}
  for word in words:
    if word not in stopwords:
      features["word({})".format(word)] = True
    return features
```

以下代码建立了一个空的 sentences 列表，用于存储所有正确标注的句子：

```
sentences = []
```

以下代码遍历 urls 字典中所有的键，并将每次遍历到的键保存到 category 变量中：

```
for category in urls.keys():
```

调用 parse() 函数从网上下载一个源，并将结果保存到 feedmap[category] 变量中：

```
feedmap[category] = feedparser.parse(urls[category])
```

使用 Python 内置的 print() 函数，将当前正在下载的 url 地址显示在屏幕上：

```
print("downloading {}".format(urls[category]))
```

遍历所有的 RSS 条目，并将每次遍历到的条目保存至 entry 变量中：

```
for entry in feedmap[category]['entries']:
```

将 RSS 源正文的摘要（summary)(新闻文本）保存到 data 变量中：

```
data = entry['summary']
```

以下代码将摘要数据按照空格分词，分词结果可传递给 NLTK 库进行特征提取：

```
words = data.split()
```

将当前 RSS 源条目中的所有单词及其对应类别以元组形式存储：

```
sentences.append((category, words))
```

提取出 sentences 中数据的所有特征，并将它们保存到 featuresets 变量中，然后调用 shuffle() 函数将列表中的所有元素随机分配到算法中：

```
featuresets = [(featureExtractor(words), category) for category, words in
sentences]
random.shuffle(featuresets)
```

建立两个数据集，其中一个是 trainset，另一个是 testset，分别用于训练和测试：

```
total = len(featuresets)
off = int(total/2)
trainset = featuresets[off:]
testset = featuresets[:off]
```

调用 NaiveBayesClassifier 模块的 train() 函数构造一个 classifier（分类器）：

```
classifier = nltk.NaiveBayesClassifier.train(trainset)
```

打印出 classifier 在 testset 上分类结果的准确率：

```
print(nltk.classify.accuracy(classifier, testset))
```

使用 classifier 类的内置函数，找出该数据的有用特征，并打印出来：

```
classifier.show_most_informative_features(5)
```

以下代码从 nfl RSS 条目中选取 4 个样本数据，并尝试根据文档标题（title）进行分类（注意，之前我们已经尝试根据摘要进行分类）：

```
for (i, entry) in enumerate(feedmap['nfl']['entries']):
  if i < 4:
    features = featureExtractor(entry['title'].split())
    category = classifier.classify(features)
    print('{} -> {}'.format(category, entry['title']))
```

7.7　利用上下文进行词性标注

在之前的章节中，我们已经学习了如何编写一个基于正则表达式的词性标注器，那时我们借助了诸如“-ed”和“-ing”等单词的后缀来判断一个词的词性。但是在英语中，同一个单词可能拥有两个词性，那么判断这类单词的词性就要依赖于上下文了。

例如，单词 address 既有名词词性又有动词词性。判断它的词性需要依赖于它所处的上下文。例句如下：

```
"What is your address when you're in Bangalore?"
"the president's address on the state of the economy."
```

在本节中，我们将借助特征提取的思想，来编写一个针对句子词性标注的工具。

7.7.1　准备工作

为了完成本节的任务，你的计算机需要安装 Python 和 NLTK 库。

7.7.2　如何实现

1. 打开 Atom 编辑器（或者你常用的程序编辑器）。
2. 创建一个新文件，命名为 ContextTagger.py。
3. 输入以下源代码：

```
ContextTagger.py
1  import nltk
2  sentences = [
3      "What is your address when you're in Bangalore?",
4      "the president's address on the state of the economy.",
5      "He addressed his remarks to the lawyers in the audience.",
6      "In order to address an assembly, we should be ready",
7      "He laughed inwardly at the scene.",
8      "After all the advance publicity, the prizefight turned out to be a laugh.",
9      "We can learn to laugh a little at even our most serious foibles."
10 ]
11 def getSentenceWords():
12     sentwords = []
13     for sentence in sentences:
14         words = nltk.pos_tag(nltk.word_tokenize(sentence))
15         sentwords.append(words)
16     return sentwords
17 def noContextTagger():
18     tagger = nltk.UnigramTagger(getSentenceWords())
19     print(tagger.tag('the little remarks towards assembly are
 .     laughable'.split()))
20 def withContextTagger():
21     def wordFeatures(words, wordPosInSentence):
22         # extract all the ing forms etc
23         endFeatures = {
24             'last(1)': words[wordPosInSentence][-1],
25             'last(2)': words[wordPosInSentence][-2:],
26             'last(3)': words[wordPosInSentence][-3:],
27         }
28         # use previous word to determine if the current word is verb or noun
29         if wordPosInSentence > 1:
30             endFeatures['prev'] = words[wordPosInSentence - 1]
31         else:
32             endFeatures['prev'] = '|NONE|'
33         return endFeatures
34     allsentences = getSentenceWords()
35     featureddata = []
36     for sentence in allsentences:
37         untaggedSentence = nltk.tag.untag(sentence)
38         featuredsentence = [(wordFeatures(untaggedSentence, index), tag) for
 .         index, (word, tag) in enumerate(sentence)]
39         featureddata.extend(featuredsentence)
40     breakup = int(len(featureddata) * 0.5)
41     traindata = featureddata[breakup:]
42     testdata = featureddata[:breakup]
43     classifier = nltk.NaiveBayesClassifier.train(traindata)
44     print("Accuracy of the classifier :
 .     {}".format(nltk.classify.accuracy(classifier, testdata)))
45
46 noContextTagger()
47 withContextTagger()
48
```

4. 保存文件。

5. 使用 Python 编译器运行程序。

6. 你将看到如下输出：

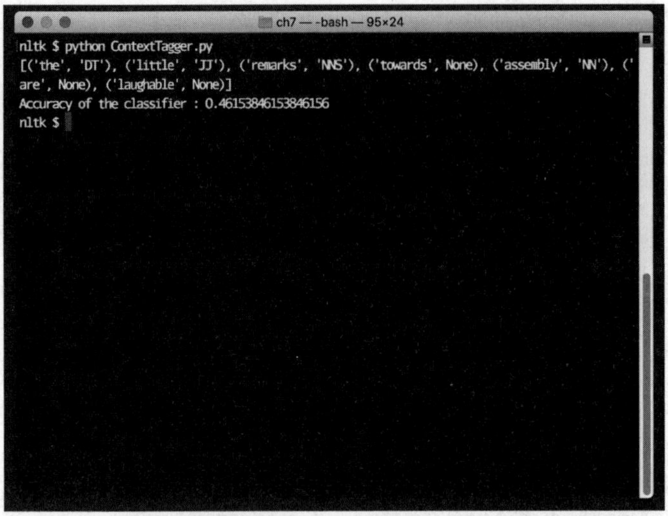

```
ch7 — -bash — 95×24
nltk $ python ContextTagger.py
[('the', 'DT'), ('little', 'JJ'), ('remarks', 'NNS'), ('towards', None), ('assembly', 'NN'), ('
are', None), ('laughable', None)]
Accuracy of the classifier : 0.46153846153846156
nltk $
```

7.7.3　工作原理

我们来理解代码是如何实现的。首先导入 NLTK 库：

```
import nltk
```

选出一些例句，这些句子都包含了具有双词性的单词，比如 address、laugh 等，并将这些句子以字符串的形式保存至 sentences 列表中。代码如下：

```
sentences = [
  "What is your address when you're in Bangalore?",
  "the president's address on the state of the economy.",
  "He addressed his remarks to the lawyers in the audience.",
  "In order to address an assembly, we should be ready",
  "He laughed inwardly at the scene.",
  "After all the advance publicity, the prizefight turned out to be a
laugh.",
  "We can learn to laugh a little at even our most serious foibles."
]
```

以下代码定义了一个函数，它的输入是 sentences 列表，然后遍历列表中的元素，最后返回一个二维列表。内部列表中包含单词及其对应的词性。代码如下：

```
def getSentenceWords():
  sentwords = []
  for sentence in sentences:
    words = nltk.pos_tag(nltk.word_tokenize(sentence))
    sentwords.append(words)
    return sentwords
```

为了衡量词性标注的质量，我们构造了一个基准系统。其中，UnigramTagger 通过当前词就能判定词性。它可以对样例数据进行学习，但是与 NLTK 库中内置标注器相比，这个标注器的效果不太理想，仅用于比较使用。代码如下：

```
def noContextTagger():
  tagger = nltk.UnigramTagger(getSentenceWords())
  print(tagger.tag('the little remarks towards assembly are
laughable'.split()))
```

以下代码定义了一个新函数，命名为 withContextTagger():

```
def withContextTagger():
```

下面这个函数的功能是在给定文本上提取特征，并返回一个字典。该字典包含当前词最后 3 个字符的信息和前一个单词的信息：

```
def wordFeatures(words, wordPosInSentence):
  # extract all the ing forms etc
  endFeatures = {
    'last(1)': words[wordPosInSentence][-1],
    'last(2)': words[wordPosInSentence][-2:],
    'last(3)': words[wordPosInSentence][-3:],
  }
  # use previous word to determine if the current word is verb or noun
  if wordPosInSentence > 1:
    endFeatures['prev'] = words[wordPosInSentence - 1]
  else:
    endFeatures['prev'] = '|NONE|'
    return endFeatures
```

建立一个 featureddata 列表，表中的元素为元组，元组中包含特征信息（featurelist）和标记（tag）两项，这两项将用于后续的朴素贝叶斯分类任务：

```
allsentences = getSentenceWords()
featureddata = []
for sentence in allsentences:
  untaggedSentence = nltk.tag.untag(sentence)
  featuredsentence = [(wordFeatures(untaggedSentence, index), tag) for
index, (word, tag) in enumerate(sentence)]
  featureddata.extend(featuredsentence)
```

提取 50% 的特征数据用于训练，50% 用于分类器的测试：

```
breakup = int(len(featureddata) * 0.5)
traindata = featureddata[breakup:]
testdata = featureddata[:breakup]
```

以下代码利用训练数据来生成分类器（classifier）。代码如下：

```
classifier = nltk.NaiveBayesClassifier.train(traindata)
```

以下代码打印出分类器在 testdata 上的分类结果准确率：

```
print("Accuracy of the classifier :
{}".format(nltk.classify.accuracy(classifier, testdata)))
```

以下两个函数打印出之前编写的两个分类器的分类结果：

```
noContextTagger()
withContextTagger()
```

CHAPTER 8

第8章 高阶自然语言处理实践

8.1 引言

目前为止，我们已经学习了如何处理输入文本、词性识别以及重要信息（命名实体）的抽取。同时我们也学习了计算机科学中的一些概念，如文法、句法分析器等。在本章中，我们将探讨**自然语言处理**中一些更高阶的任务，这些任务需要采用多种技术手段才能实现。

8.2 创建一条自然语言处理管道

在计算机领域，一条管道可以被看作是一个多阶段的数据流系统，其中一个组件的输出被视为另一个组件的输入。

管道具备以下特点：

- 数据始终从一个组件流向另一个组件
- 组件是一个只考虑输入和输出数据的黑盒

一条定义良好的管道应该考虑以下事情：

- 流经每个组件的输入数据的格式
- 流经每个组件的输出数据的格式
- 通过调整数据流入和流出的速度，确保组件之间的数据流量得到控制

例如，如果你熟悉 Unix/Linux 系统，并且了解 shell 上相关的操作，你一定知道“|”运算符，它就是 shell 中数据管道的抽象。我们可以利用“|”运算符在 Unix shell 中创建管道。

为了能更轻松地理解，我们举一个 Unix 系统中的例子：如何查找给定目录中文件的数量？

为了解决这个问题，我们需要做以下事情：

- 我们需要一个组件（或 Unix 环境中的一个命令）来读取目录并列出其中的所有文件
- 我们需要另一个组件（或 Unix 环境中的一个命令）来读取这些行并打印行数

所以，我们用以下两个命令来完成上面的任务：

- ls 命令
- wc 命令

如果可以建立一条管道，将 ls 的输出内容传递给 wc，我们就解决这个问题了。

在 Unix 命令中，ls -l | wc -l 就是一条能对当前目录中文件进行计数的管道。

有了以上知识，我们知道 NLP 管道需要具备以下功能：

- 采集输入数据
- 对输入数据进行分词
- 识别输入数据中单词的词性（POS）
- 从单词中抽取命名实体
- 识别命名实体之间的关系

在本节中，我们尽可能尝试简单地构建一条管道。它从远程的 RSS 源获取数据，然后将其中识别到的命名实体输出到对应的文件中去。

8.2.1　准备工作

你需要安装 Python 以及 NLTK、queue、feedparser 和 uuid 库。

8.2.2　如何实现

1. 打开 Atom 编辑器（或者你常用的程序编辑器）。
2. 创建一个新文件，命名为 PipelineQ.py。
3. 输入以下源代码：

PipelineQ.py

```
import nltk
import threading
import queue
import feedparser
import uuid

threads = []
queues = [queue.Queue(), queue.Queue()]

def extractWords():
    url = 'https://timesofindia.indiatimes.com/rssfeeds/1081479906.cms'
    feed = feedparser.parse(url)
    for entry in feed['entries'][:5]:
        text = entry['title']
        if 'ex' in text:
            continue
        words = nltk.word_tokenize(text)
```

```
        data = {'uuid': uuid.uuid4(), 'input': words}
        queues[0].put(data, True)
        print(">> {} : {}".format(data['uuid'], text))

def extractPOS():
    while True:
        if queues[0].empty():
            break
        else:
            data = queues[0].get()
            words = data['input']
            postags = nltk.pos_tag(words)
            queues[0].task_done()
            queues[1].put({'uuid': data['uuid'], 'input': postags}, True)

def extractNE():
    while True:
        if queues[1].empty():
            break
        else:
            data = queues[1].get()
            postags = data['input']
            queues[1].task_done()
            chunks = nltk.ne_chunk(postags, binary=False)
            print("  << {} : ".format(data['uuid']), end = '')
            for path in chunks:
                try:
                    label = path.label()
                    print(path, end=', ')
                except:
                    pass
            print()

def runProgram():
    e = threading.Thread(target=extractWords())
    e.start()
    threads.append(e)

    p = threading.Thread(target=extractPOS())
    p.start()
    threads.append(p)

    n = threading.Thread(target=extractNE())
    n.start()
    threads.append(n)

    queues[0].join()
    queues[1].join()

    for t in threads:
        t.join()

if __name__ == '__main__':
    runProgram()

```

4. 保存文件。
5. 使用 Python 编译器运行程序。
6. 你将看到以下输出：

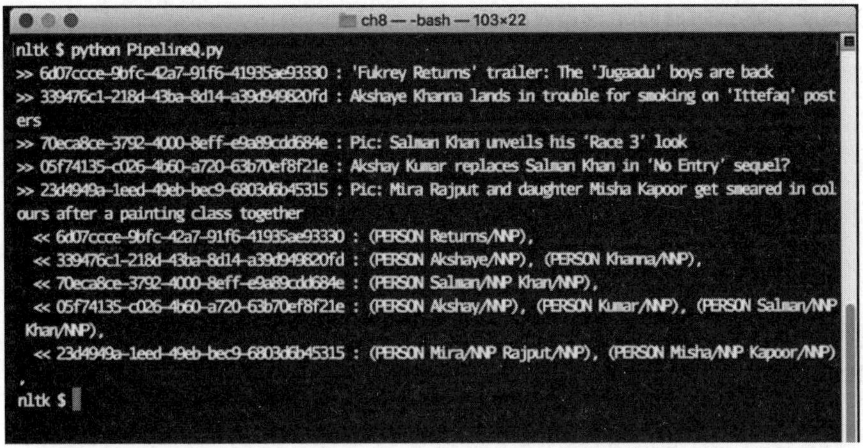

```
ch8 — -bash — 103×22
nltk $ python PipelineQ.py
>> 6d07ccce-9bfc-42a7-91f6-41935ae93330 : 'Fukrey Returns' trailer: The 'Jugaadu' boys are back
>> 339476c1-218d-43ba-8d14-a39d949820fd : Akshaye Khanna lands in trouble for smoking on 'Ittefaq' post
ers
>> 70eca8ce-3792-4000-8eff-e9a89cdd684e : Pic: Salman Khan unveils his 'Race 3' look
>> 05f74135-c026-4b60-a720-63b70ef8f21e : Akshay Kumar replaces Salman Khan in 'No Entry' sequel?
>> 23d4949a-1eed-49eb-bec9-6803d6b45315 : Pic: Mira Rajput and daughter Misha Kapoor get smeared in col
ours after a painting class together
  << 6d07ccce-9bfc-42a7-91f6-41935ae93330 : (PERSON Returns/NNP),
  << 339476c1-218d-43ba-8d14-a39d949820fd : (PERSON Akshaye/NNP), (PERSON Khanna/NNP),
  << 70eca8ce-3792-4000-8eff-e9a89cdd684e : (PERSON Salman/NNP Khan/NNP),
  << 05f74135-c026-4b60-a720-63b70ef8f21e : (PERSON Akshay/NNP), (PERSON Kumar/NNP), (PERSON Salman/NNP
 Khan/NNP),
  << 23d4949a-1eed-49eb-bec9-6803d6b45315 : (PERSON Mira/NNP Rajput/NNP), (PERSON Misha/NNP Kapoor/NNP)
,
nltk $
```

8.2.3　工作原理

接下来我们来了解管道的构建过程：

```
import nltk
import threading
import queue
import feedparser
import uuid
```

以上 5 行代码将 5 个 Python 库导入到了当前程序中：

- NLTK：自然语言工具包
- threading：用于在单个程序中创建轻量级任务的线程库
- queue：可以在多线程程序中使用的队列库
- feedparser:：RSS 源解析库
- uuid：基于 RFC-4122 的 uuid 版本 1、3、4、5 的生成库

```
threads = []
```

创建一个新的空列表来跟踪程序中的所有线程：

```
queues = [queue.Queue(), queue.Queue()]
```

这条代码创建了一个包含两个队列对象的列表对象 queues。

为什么我们需要两个队列：

- 第 1 个队列用于存储分词后的句子
- 第 2 个队列用于存储所有标注过词性的单词

以下代码定义了一个新的函数 extractWords()，它将从互联网上读取 RSS 源的一个文本并存储其中的单词以及该文本的唯一标识符：

```
def extractWords():
```

以下代码定义了来自印度时报网站（娱乐新闻）的示例 URL：

```
url = 'https://timesofindia.indiatimes.com/rssfeeds/1081479906.cms'
```

以下代码调用了 feedparser 库中的 parse() 函数。parse() 函数会下载 URL 的内容并将其转换为新闻项目的列表。每个新闻条目都是一个由标题和摘要组成的字典：

```
feed = feedparser.parse(url)
```

从 RSS 源中获取前 5 个条目，并将当前条目存储到名为 entry 的变量中：

```
for entry in feed['entries'][:5]:
```

将当前 RSS 源条目中的标题存储到名为 text 的变量中：

```
text = entry['title']
```

以下代码会跳过包含敏感词的标题。由于我们从互联网上获取数据，所以我们要保证获得的数据是过滤过的安全数据：

```
if 'ex' in text:
  continue
```

使用 word_tokenize() 函数对输入文本进行分词，并将结果存储到名为 words 的变量中：

```
words = nltk.word_tokenize(text)
```

用两个键值对创建一个名为 data 的字典，用其分别存储 UUID 和输入的单词：

```
data = {'uuid': uuid.uuid4(), 'input': words}
```

以下代码将上面的字典存储在第一个队列，即 queues[0] 中。第二个参数设置为 true，表示如果队列已满，则暂停该线程：

```
queues[0].put(data, True)
```

一条设计良好的管道应该根据组件的计算能力来控制数据的输入和输出，否则，整条管道将会崩溃。以下代码打印输出当前的 RSS 条目及其唯一的 ID：

```
print(">> {} : {}".format(data['uuid'], text))
```

以下代码定义了一个名为 extractPOS() 的函数，该函数从第一个队列中读取数据并处理数据，并将这些单词的词性保存在第二个队列中：

```
def extractPOS():
```

接下来是一个无限循环：

```
while True:
```

以下代码检查第一个队列是否为空，如果为空，则停止处理：

```
if queues[0].empty():
  break
```

为了增加程序的鲁棒性，可以传递第一个队列的反馈信息，这里作为一个练习留给读者。以下是 else 部分，表示第一个队列有一些数据：

```
else:
```

按照先进先出（FIFO）顺序，取出队列中的第一个元素：

```
data = queues[0].get()
```

识别出单词的词性：

```
words = data['input']
postags = nltk.pos_tag(words)
```

更新第一个队列，表示我们已经处理完毕由此线程获取的条目：

```
queues[0].task_done()
```

将标注词性的单词列表存储在第二个队列中，以便管道中的下一个组件能够进行处理。在这里，我们将 true 作为第二个参数，确保线程在队列空间已满时执行等待：

```
queues[1].put({'uuid': data['uuid'], 'input': postags}, True)
```

以下代码定义了一个新的函数 extractNE()，它从第二个队列中读取数据，即处理标注词性后的单词，并在屏幕上打印出命名实体：

```
def extractNE():
```

接下来是一个无限循环：

```
while True:
```

如果第二个队列为空，则终止循环：

```
if queues[1].empty():
  break
```

以下代码将从第二个队列中取出一个元素，并将其存入名为 data 的变量中：

```
else:
  data = queues[1].get()
```

以下代码表示我们已经处理完毕从第二个队列获取到的元素：

```
postags = data['input']
queues[1].task_done()
```

以下代码将 postags 变量中的命名实体抽取出来并存入名为 chunks 的变量中：

```
chunks = nltk.ne_chunk(postags, binary=False)

print("  << {} : ".format(data['uuid']), end = '')
  for path in chunks:
    try:
      label = path.label()
      print(path, end=', ')
```

```
        except:
            pass
    print()
```

以上代码完成以下工作：

- 打印 data 字典中的 UUID
- 遍历所有识别出来的组块
- 使用 try/except 块，因为并不是树中的所有元素都有 label() 函数（当没有找到组块时，它们是元组，也就没有 label 函数）
- 最后，调用 print() 函数，在屏幕上输出换行符

以下代码定义了一个新函数 runProgram，使用线程进行管道的设置：

```
def runProgram():
```

以下三行代码创建了一个新的线程来运行 extractWords() 函数，启动该线程并将其添加到名为 threads 的列表中：

```
e = threading.Thread(target=extractWords())
e.start()
threads.append(e)
```

以下代码创建了一个新的线程来运行 extractPOS() 函数，启动该线程并将其添加到名为 threads 的列表中：

```
p = threading.Thread(target=extractPOS())
p.start()
threads.append(p)
```

以下代码创建了一个新的线程来运行 extractNE() 函数，启动该线程并将其添加到名为 threads 的列表中：

```
n = threading.Thread(target=extractNE())
n.start()
threads.append(n)
```

以下两行代码将在所有工作处理完毕后释放分配给 queues 的资源：

```
queues[0].join()
queues[1].join()
```

以下两行代码遍历线程列表，将当前的线程对象存储在变量 t 中，t 调用 join() 函数来标记线程的完成，并释放分配给线程的资源：

```
for t in threads:
    t.join()
```

以下是程序中主线程运行时调用的代码段。调用 runProgram() 函数来启动整个管道：

```
if __name__ == '__main__':
    runProgram()
```

8.3 解决文本相似度问题

文本相似度问题就是计算两个给定文本之间的相似程度。目前，我们可以从多个维度出发来判别文本之间的相似度：

- 情感（Sentiment/emotion）维度
- 感官（Sense）维度
- 特定词的出现

当前已经有许多可用的算法，但是它们在算法复杂度、资源需求，以及可处理数据量等方面都有很大差异。

在本节中，我们将使用 TF-IDF 算法来解决相似度问题。首先我们来了解基本概念：

- **词频（Term frequency，TF）**：这个概念试图计算给定文档中词语的相对重要性（频率）。

由于我们讨论的是相对重要性，因此通常将文档中所有词的出现频率正则化以计算一个词的 TF 值。

- **逆文本频率（Inverse Document Frequency，IDF）**：这个技术赋予那些经常出现的单词（如 a、the 等）更低的权重值（与那些少见的单词比较）。

由于 TF 和 IDF 值都计算为数字（小数），所以我们要对每个文档的每个词的这两个值进行乘法运算，并构建 N 个 M 维向量（这里 N 是文档总数，M 是所有文档的去重词汇量）。

当我们有了这些向量，就可以利用下面的公式对每个向量求余弦相似度：

$$similarity = \cos(\theta) = \frac{\boldsymbol{A} \cdot \boldsymbol{B}}{\|\boldsymbol{A}\|_2 \|\boldsymbol{B}\|_2} = \frac{\sum_{i=1}^{n} \boldsymbol{A}_i \boldsymbol{B}_i}{\sqrt{\sum_{i=1}^{n} \boldsymbol{A}_i^2} \sqrt{\sum_{i=1}^{n} \boldsymbol{B}_i^2}},$$

其中 $\boldsymbol{A}_i$ 和 $\boldsymbol{B}_i$ 分别是向量 $\boldsymbol{A}$ 和 $\boldsymbol{B}$ 的分量。

8.3.1 准备工作

你需要安装 Python，然后安装 NLTK 及 scikit 库。本项任务对读者的数学基础有一定要求。

8.3.2 如何实现

1. 打开 Atom 编辑器（或者你常用的程序编辑器）。
2. 创建一个文件，命名为 Similarity.py。
3. 输入以下源代码：

Similarity.py

```
1  import nltk
2  import math
3  from sklearn.feature_extraction.text import TfidfVectorizer
```

```python
from sklearn.metrics.pairwise import cosine_similarity

class TextSimilarityExample:
    def __init__(self):
        self.statements = [
            'ruled india',
            'Chalukyas ruled Badami',
            'So many kingdoms ruled India',
            'Lalbagh is a botanical garden in India'
        ]
    def TF(self, sentence):
        words = nltk.word_tokenize(sentence.lower())
        freq = nltk.FreqDist(words)
        dictionary = {}
        for key in freq.keys():
            norm = freq[key]/float(len(words))
            dictionary[key] = norm
        return dictionary

    def IDF(self):
        def idf(TotalNumberOfDocuments, NumberOfDocumentsWithThisWord):
            return 1.0 +
            math.log(TotalNumberOfDocuments/NumberOfDocumentsWithThisWord)
        numDocuments = len(self.statements)
        uniqueWords = {}
        idfValues = {}
        for sentence in self.statements:
            for word in nltk.word_tokenize(sentence.lower()):
                if word not in uniqueWords:
                    uniqueWords[word] = 1
                else:
                    uniqueWords[word] += 1
        for word in uniqueWords:
            idfValues[word] = idf(numDocuments, uniqueWords[word])
        return idfValues

    def TF_IDF(self, query):
        words = nltk.word_tokenize(query.lower())
        idf = self.IDF()
        vectors = {}
        for sentence in self.statements:
            tf = self.TF(sentence)
            for word in words:
                tfv = tf[word] if word in tf else 0.0
                idfv = idf[word] if word in idf else 0.0
                mul = tfv * idfv
                if word not in vectors:
                    vectors[word] = []
                vectors[word].append(mul)
        return vectors

    def displayVectors(self, vectors):
        print(self.statements)
        for word in vectors:
            print("{} -> {}".format(word, vectors[word]))

    def cosineSimilarity(self):
        vec = TfidfVectorizer()
        matrix = vec.fit_transform(self.statements)
```

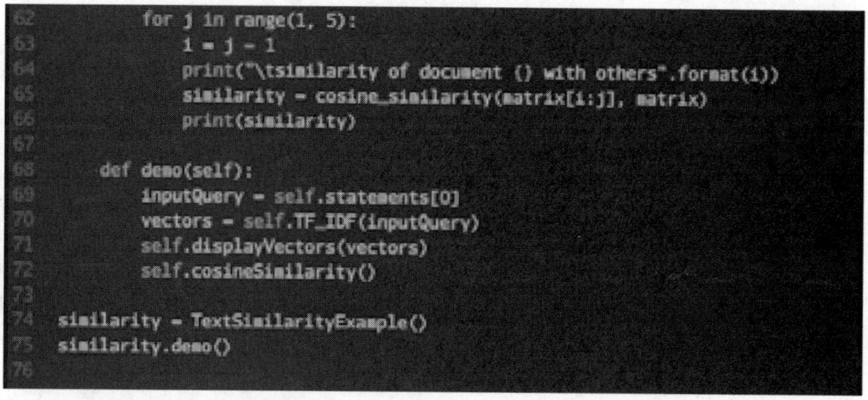

```
62          for j in range(1, 5):
63              i = j - 1
64              print("\tsimilarity of document {} with others".format(i))
65              similarity = cosine_similarity(matrix[i:j], matrix)
66              print(similarity)
67
68      def demo(self):
69          inputQuery = self.statements[0]
70          vectors = self.TF_IDF(inputQuery)
71          self.displayVectors(vectors)
72          self.cosineSimilarity()
73
74  similarity = TextSimilarityExample()
75  similarity.demo()
76
```

4. 保存文件。
5. 使用 Python 编译器运行程序。
6. 你将看到如下输出：

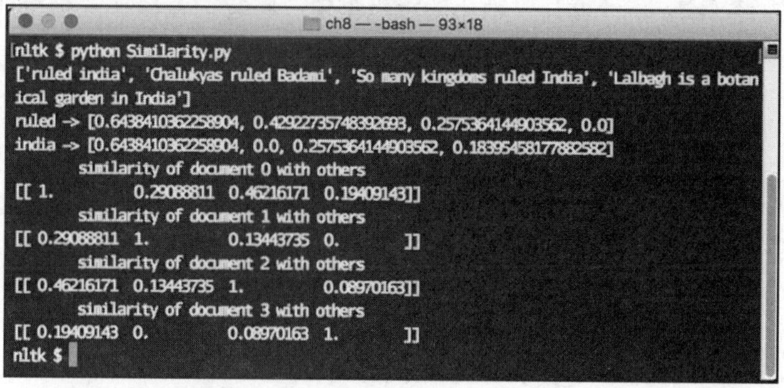

```
ch8 — -bash — 93×18
nltk $ python Similarity.py
['ruled india', 'Chalukyas ruled Badami', 'So many kingdoms ruled India', 'Lalbagh is a botan
ical garden in India']
ruled -> [0.643841036225804, 0.42922735748392693, 0.2575364144903562, 0.0]
india -> [0.643841036225804, 0.0, 0.2575364144903562, 0.18395458177882582]
        similarity of document 0 with others
[[ 1.          0.29088811  0.46216171  0.19409143]]
        similarity of document 1 with others
[[ 0.29088811  1.          0.13443735  0.        ]]
        similarity of document 2 with others
[[ 0.46216171  0.13443735  1.          0.08970163]]
        similarity of document 3 with others
[[ 0.19409143  0.          0.08970163  1.        ]]
nltk $
```

8.3.3　工作原理

我们来看如何解决文本相似度问题。下面的代码导入该项目所需的包：

```
import nltk
import math
from sklearn.feature_extraction.text import TfidfVectorizer
from sklearn.metrics.pairwise import cosine_similarity
```

现在我们定义一个新的类，类名为 TextSimilarityExample：

```
class TextSimilarityExample:
```

以下声明语句定义了一个新的类构造器：

```
def __init__(self):
```

下面的代码定义了我们需要计算相似度的例句：

```
self.statements = [
  'ruled india',
```

```
    'Chalukyas ruled Badami',
    'So many kingdoms ruled India',
    'Lalbagh is a botanical garden in India'
]
```

下面的代码定义了 TF 函数来计算句子中所有单词的 TF 值：

```
def TF(self, sentence):
    words = nltk.word_tokenize(sentence.lower())
    freq = nltk.FreqDist(words)
    dictionary = {}
    for key in freq.keys():
        norm = freq[key]/float(len(words))
        dictionary[key] = norm
    return dictionary
```

该函数功能如下：

- 把文本转换成小写字母并提取所有的单词；
- 使用 NLTK 内置的 FreqDist 函数，计算这些词的词频分布；
- 遍历字典中所有的键（key），计算其正则化的浮点型值，将它们存储在一个字典中；
- 返回该字典，其中包含了句子中所有单词正则化后的频率值。

以下代码定义了一个 IDF 函数，它可以计算文档中所有单词的 IDF 值。

```
    def IDF(self):
        def idf(TotalNumberOfDocuments, NumberOfDocumentsWithThisWord):
            return 1.0 +
math.log(TotalNumberOfDocuments/NumberOfDocumentsWithThisWord)
        numDocuments = len(self.statements)
        uniqueWords = {}
        idfValues = {}
        for sentence in self.statements:
            for word in nltk.word_tokenize(sentence.lower()):
                if word not in uniqueWords:
                    uniqueWords[word] = 1
                else:
                    uniqueWords[word] += 1
        for word in uniqueWords:
            idfValues[word] = idf(numDocuments, uniqueWords[word])
        return idfValues
```

该步骤做了以下工作：

- 定义了一个名为 idf() 的局部函数，该函数可以计算给定单词的 IDF 值；
- 遍历所有的句子并将它们都转换成小写字母；
- 计算每个单词在所有文档中出现的次数；
- 计算所有单词的 IDF 值并返回一个包含这些 IDF 值的字典。

以下代码定义了一个 TF_IDF(TF 值与 IDF 值相乘）函数，通过给定搜索字符串（search string）查找所有文档：

```
def TF_IDF(self, query):
    words = nltk.word_tokenize(query.lower())
```

```
        idf = self.IDF()
        vectors = {}
        for sentence in self.statements:
            tf = self.TF(sentence)
            for word in words:
                tfv = tf[word] if word in tf else 0.0
                idfv = idf[word] if word in idf else 0.0
                mul = tfv * idfv
                if word not in vectors:
                    vectors[word] = []
                vectors[word].append(mul)
        return vectors
```

现在我们来看以上代码做了哪些工作：

- 将这些字符串进行分词；
- 在 self.statements 变量中为所有句子构造一个 IDF() 函数；
- 遍历所有的句子并计算当前句子中所有单词的 TF 值；
- 仅针对输入搜索字符串所包含的单词，计算 tf*idf 值来构建每个文本的向量；
- 返回搜索字符串中每个单词的向量列表。

以下函数在屏幕上显示出向量内容：

```
    def displayVectors(self, vectors):
        print(self.statements)
        for word in vectors:
            print("{} -> {}".format(word, vectors[word]))
```

正如我们开始讨论时介绍的，为了得到相似度我们需要计算输入向量的余弦相似度。我们可以编写代码进行这些数学计算。但是这次，我们尝试使用 scikit 库来完成所有的这些计算：

```
    def cosineSimilarity(self):
        vec = TfidfVectorizer()
        matrix = vec.fit_transform(self.statements)
        for j in range(1, 5):
            i = j - 1
            print("\tsimilarity of document {} with others".format(i))
            similarity = cosine_similarity(matrix[i:j], matrix)
            print(similarity)
```

在之前的函数中，我们学习了如何计算 TF 和 IDF 的值，并且最终获得了所有文档的 TF × IDF 值。

让我们看看以上的代码做了什么工作：

- 定义了一个新函数：cosineSimilarity()；
- 创建了一个新的向量对象；
- 使用 fit_transform() 函数计算我们感兴趣的所有文本的 TF-IDF 值矩阵；
- 接下来我们将每个文档与所有的其他文档作比较，计算它们之间的相似度。

以下代码定义了 demo() 函数，用来运行我们已定义的所有函数：

```
    def demo(self):
        inputQuery = self.statements[0]
        vectors = self.TF_IDF(inputQuery)
        self.displayVectors(vectors)
        self.cosineSimilarity()
```

现在看看以上代码做了什么工作：

- 把第一条句子作为我们的输入查询（query）；
- 利用自己编写的 TF_IDF() 函数建立向量；
- 在屏幕上显示所有句子的 TF × IDF 向量；
- 调用 scikit 库的 cosineSimilarity() 函数，计算输入查询与所有句子的余弦相似度并打印出来。

以下代码中，我们为 TextSimilarityExample() 类创建了一个新的对象，然后调用 demo() 函数。

```
similarity = TextSimilarityExample()
similarity.demo()
```

8.4 主题识别

在之前章节中，我们学习了如何进行文本分类。初学者可能会认为文本分类和主题识别是一样的，但是它们之间存在细微的差别。

主题识别是发现输入文本集合中存在的主题的过程。这些主题可能是给定文本中只出现一次的一些单词。

例如，当我们阅读到提及萨辛－泰杜尔卡（Sachin Tendulkar）、得分（score）和胜利（win）的相关文本时，我们可以知道这是在描述板球。当然这种判断有时也可能是错误的。

为了发现给定输入文本的所有主题，我们使用潜在狄利克雷分布（Latent Dirichlet allocation，LDA）算法（我们也可以使用 TF-IDF 算法，但在前面的小节中我们已经探讨过它，现在我们来了解 LDA 如何识别主题）。

8.4.1 准备工作

首先你需要安装 Python，以及 NLTK、gensim 和 feedparser 库。

8.4.2 如何实现

1. 打开 Atom 编辑器（或者你常用的程序编辑器）。
2. 创建一个名为 IdentifyingTopic.py 的文件。
3. 输入以下源代码：

```
IdentifyingTopic.py
1  from nltk.tokenize import RegexpTokenizer
```

```
from nltk.corpus import stopwords
from gensim import corpora, models
import nltk
import feedparser

class IdentifyingTopicExample:
    def getDocuments(self):
        url = 'https://sports.yahoo.com/mlb/rss.xml'
        feed = feedparser.parse(url)
        self.documents = []
        for entry in feed['entries'][:5]:
            text = entry['summary']
            if 'ex' in text:
                continue
            self.documents.append(text)
            print("-- {}".format(text))
        print("INFO: Fetching documents from {} completed".format(url))

    def cleanDocuments(self):
        tokenizer = RegexpTokenizer(r'[a-zA-Z]+')
        en_stop = set(stopwords.words('english'))
        self.cleaned = []
        for doc in self.documents:
            lowercase_doc = doc.lower()
            words = tokenizer.tokenize(lowercase_doc)
            non_stopped_words = [i for i in words if not i in en_stop]
            self.cleaned.append(non_stopped_words)
        print("INFO: Clearning {} documents
        completed".format(len(self.documents)))

    def doLDA(self):
        dictionary = corpora.Dictionary(self.cleaned)
        corpus = [dictionary.doc2bow(cleandoc) for cleandoc in self.cleaned]
        ldamodel = models.ldamodel.LdaModel(corpus, num_topics=2, id2word =
        dictionary)
        print(ldamodel.print_topics(num_topics=2, num_words=4))

    def run(self):
        self.getDocuments()
        self.cleanDocuments()
        self.doLDA()

if __name__ == '__main__':
    topicExample = IdentifyingTopicExample()
    topicExample.run()
```

4. 保存文件。
5. 使用 Python 编译器运行程序。
6. 你将看到以下输出：

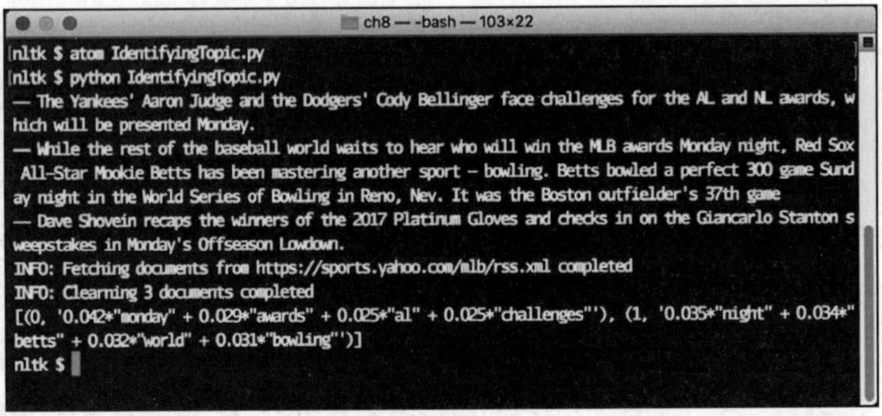

```
ch8 — -bash — 103×22
nltk $ atom IdentifyingTopic.py
nltk $ python IdentifyingTopic.py
— The Yankees' Aaron Judge and the Dodgers' Cody Bellinger face challenges for the AL and NL awards, w
hich will be presented Monday.
— While the rest of the baseball world waits to hear who will win the MLB awards Monday night, Red Sox
 All-Star Mookie Betts has been mastering another sport - bowling. Betts bowled a perfect 300 game Sund
ay night in the World Series of Bowling in Reno, Nev. It was the Boston outfielder's 37th game
— Dave Shovein recaps the winners of the 2017 Platinum Gloves and checks in on the Giancarlo Stanton s
weepstakes in Monday's Offseason Lowdown.
INFO: Fetching documents from https://sports.yahoo.com/mlb/rss.xml completed
INFO: Clearning 3 documents completed
[(0, '0.042*"monday" + 0.029*"awards" + 0.025*"al" + 0.025*"challenges"'), (1, '0.035*"night" + 0.034*"
betts" + 0.032*"world" + 0.031*"bowling"')]
nltk $
```

8.4.3 工作原理

我们来看主题识别程序是如何工作的。首先通过以下代码将必需的库导入到当前程序中：

```
from nltk.tokenize import RegexpTokenizer
from nltk.corpus import stopwords
from gensim import corpora, models
import nltk
import feedparser
```

以下代码定义了一个新的类 IdentifyingTopicExample：

```
class IdentifyingTopicExample:
```

以下代码定义了一个新的函数 getDocuments()，它的功能是利用 feedparser 函数从互联网上下载新的文档：

```
def getDocuments(self):
```

以下代码通过给定的 URL 下载所有的文档，将其分析后的字典列表存储到 feed 变量中：

```
url = 'https://sports.yahoo.com/mlb/rss.xml'
feed = feedparser.parse(url)
```

为了记录将要进一步分析的文档，我们首先将列表清空：

```
self.documents = []
```

从 feed 变量中取出前 5 个文档并将这些新闻条目存储到名为 entry 的变量中：

```
for entry in feed['entries'][:5]:
```

将新闻摘要存储到名为 text 的变量中：

```
text = entry['summary']
```

跳过包含敏感词汇的新闻文本：

```
if 'ex' in text:
    continue
```

将文本保存在 documents 变量中：

```
self.documents.append(text)
```

在屏幕上显示出当前文档：

```
print("-- {}".format(text))
```

以下代码向用户显示了我们已经从给定的 url 中收集到的 N 个文档的提示消息：

```
print("INFO: Fetching documents from {} completed".format(url))
```

以下代码定义了一个新的函数 cleanDocuments()，它的作用是清洗输入文本（由于我们从因特网上下载数据，它可能包含任何类型的数据）。

```
def cleanDocuments(self):
```

这里我们只对由英文字母组成的单词感兴趣。所以，对文本进行分词后的分词结果被定义成由从 a 到 z 及从 A 到 Z 的字母组成。这样可以确保标点符号和其他无关紧要的数据不会进入到下一步的处理过程中：

```
tokenizer = RegexpTokenizer(r'[a-zA-Z]+')
```

将英语的停用词存储到 en_stop 变量中：

```
en_stop = set(stopwords.words('english'))
```

定义一个空的列表集合 cleaned，用来存储所有被清洗且分词后的文档：

```
self.cleaned = []
```

调用 getDocuments() 函数，遍历我们收集的所有文档：

```
for doc in self.documents:
```

由于大小写敏感问题，以下代码将文档统一转换成小写字母，避免同一单词采用不同的处理方式：

```
lowercase_doc = doc.lower()
```

对句子进行分词，输出的单词列表存储在 words 变量中：

```
words = tokenizer.tokenize(lowercase_doc)
```

去除所有英语停用词，将句子中其他单词存储在名为 non_stopped_words 的变量中：

```
non_stopped_words = [i for i in words if i not in en_stop]
```

将分词及清洗后的句子保存在 self.cleaned 变量（类成员）中：

```
self.cleaned.append(non_stopped_words)
```

向用户显示我们已完成文档清洗：

```
        print("INFO: Cleaning {} documents
completed".format(len(self.documents)))
```

以下代码定义了一个新的函数 doLDA，它的功能是在清洗后的文档上执行 LDA 分析：

```
def doLDA(self):
```

在处理这些清洗后的文档前，我们将这些文档创建一个字典：

```
dictionary = corpora.Dictionary(self.cleaned)
```

由每一条清洗后的句子，以词袋（a bag of words）形式定义 corpus 变量：

```
corpus = [dictionary.doc2bow(cleandoc) for cleandoc in self.cleaned]
```

在 corpus 上创建一个模型，将主题数量设置为 2，并用 id2word 参数设置词典的大小/映射情况（size/mapping）：

```
ldamodel = models.ldamodel.LdaModel(corpus, num_topics=2, id2word =
dictionary)
```

在屏幕上打印出两个主题，每个主题应该包含 4 个单词：

```
print(ldamodel.print_topics(num_topics=2, num_words=4))
```

以下函数按顺序执行了所有步骤：

```
def run(self):
    self.getDocuments()
    self.cleanDocuments()
    self.doLDA()
```

当主（main）程序调用上述程序时，从 IdentifyingTopicExample() 类创建一个新的 topicExample 对象，并在该对象上调用 run() 函数。

```
if __name__ == '__main__':
    topicExample = IdentifyingTopicExample()
    topicExample.run()
```

8.5 文本摘要

在这个信息爆炸的时代，通过出版物/文本的形式可以获得大量信息。由于我们不可能使用所有的数据，所以为了便捷使用这些数据，我们一直在开发一些算法将大文本简化为一个容易得到的摘要（或要点）。这种做法可以节约时间，并使互联网工作处理起来更容易。在本节中，我们将使用 gensim 库，采用内置的 TextRank 算法实现文本摘要。(https://web.eecs.umich.edu/~mihalcea/ papers/mihalcea.emnlp04.pdf).

8.5.1 准备工作

安装 Python，以及 bs4 库和 gensim 数据库。

8.5.2 如何实现

1. 打开 Atom 编辑器（或者你常用的程序编辑器）。
2. 创建一个新文档，命名为 Summarize.py。
3. 输入以下源代码：

Summarize.py

```python
from gensim.summarization import summarize
from bs4 import BeautifulSoup
import requests

#
# This recipe uses automatic computer science Paper generation tool from mit.edu
# You can generate your own paper by visiting https://pdos.csail.mit.edu/archive/scigen/
# and click generate.
#
# This example needs large amount of text that needs to be available for summary.
# So, we are using this paper generation tool and extracting the 'Introduction' section
# to do the summary analysis.
#

urls = {
    'Daff: Unproven Unification of Suffix Trees and Redundancy': 'http://scigen.csail.mit.edu/scicache/610/scimakelatex.21945.none.html',
    'CausticIslet: Exploration of Rasterization': 'http://scigen.csail.mit.edu/scicache/790/scimakelatex.1499.none.html'
}

for key in urls.keys():
    url = urls[key]
    r = requests.get(url)
    soup = BeautifulSoup(r.text, 'html.parser')
    data = soup.get_text()
    pos1 = data.find("1  Introduction") + len("1  Introduction")
    pos2 = data.find("2  Related Work")
    text = data[pos1:pos2].strip()
    print("PAPER URL: {}".format(url))
    print("TITLE: {}".format(key))
    print("GENERATED SUMMARY: {}".format(summarize(text)))
    print()

```

4. 保存文件。

5. 使用 Python 编译器运行程序。

6. 你将看到如下输出：

```
ch8 — -bash — 104×23
nltk $ python Summarize.py
PAPER URL: http://scigen.csail.mit.edu/scicache/610/scimakelatex.21945.none.html
TITLE: Daff: Unproven Unification of Suffix Trees and Redundancy
GENERATED SUMMARY: In this work we better understand how write-ahead logging  can be
applied to the understanding of consistent hashing.
frameworks emulate low-energy communication.
deployed in existing work.
To our knowledge, our work in this paper marks the first algorithm
Nevertheless, this solution is entirely useful.
Furthermore, we better understand how the memory bus  can be
applied to the study of von Neumann machines.
present a framework for Byzantine fault tolerance  (CausticIslet),
disconfirming that telephony  and von Neumann machines  are largely

PAPER URL: http://scigen.csail.mit.edu/scicache/790/scimakelatex.1499.none.html
TITLE: CausticIslet: Exploration of Rasterization
GENERATED SUMMARY: World Wide Web, which embodies the theoretical principles of theory.
To what extent can randomized algorithms  be evaluated to address
Thus, our algorithm simulates the evaluation of write-back caches
The rest of the paper proceeds as follows.
the World Wide Web  can be applied to the deployment of XML.

nltk $
```

8.5.3 工作原理

我们来了解摘要算法是如何工作的：

```
from gensim.summarization import summarize
from bs4 import BeautifulSoup
import requests
```

这 3 条代码将必需的库导入到了当前程序中：

- `gensim.summarization.summarize`：基于文本排序的摘要算法（Text-rank-based summarization）
- `bs4`：用于解析 HTML 文档的 BeautifulSoup 库
- `requests`：用于下载 HTTP 资源的库

定义一个名为 urls 的字典，键是自动生成的文章的标题，键值是该文章对应的 URL：

```
urls = {
    'Daff: Unproven Unification of Suffix Trees and Redundancy':
'http://scigen.csail.mit.edu/scicache/610/scimakelatex.21945.none.html',
    'CausticIslet: Exploration of Rasterization':
'http://scigen.csail.mit.edu/scicache/790/scimakelatex.1499.none.html'
}
```

遍历字典的所有键：

```
for key in urls.keys():
```

将当前文章的 URL 存储在名为 url 的变量中：

```
url = urls[key]
```

使用 requests 库的 get() 方法下载 url 的内容，并将返回的对象存储到变量 r 中：

```
r = requests.get(url)
```

使用 BeautifulSoup() 方法解析 r 中的文本，参数设置为 HTML 解析器，并将返回对象存储在名为 soup 的变量中：

```
soup = BeautifulSoup(r.text, 'html.parser')
```

去掉所有的 HTML 标签，将文档中的文本提取到 data 变量中：

```
data = soup.get_text()
```

找到文本引言（Introduction）的位置并跳到这个字符串的末尾，将 pos1 标记为我们想要提取的字符串的起始位置：

```
pos1 = data.find("1  Introduction") + len("1  Introduction")
```

找到文档中的第二个位置，即相关工作（Related Work）部分的开头：

```
pos2 = data.find("2  Related Work")
```

现在，提取文档中介于这两个位置之间的引言部分：

```
text = data[pos1:pos2].strip()
```

在屏幕上显示文档的 URL 和标题：

```
print("PAPER URL: {}".format(url))
print("TITLE: {}".format(key))
```

在文本上调用 summarize() 函数，该函数根据文本排序算法返回摘要文本：

```
print("GENERATED SUMMARY: {}".format(summarize(text)))
```

为提高屏幕输出的可读性，额外打印一个换行符：

```
print()
```

8.6　指代消解

在许多自然语言句子中，代词常用于替代某些重复出现的名词，以简化句子的结构。例如：

Ravi is a boy. He often donates money to the poor.

在这个例子中，有两条语句：

- Ravi 是一个男孩

- 他经常捐钱给贫穷人

当分析第二条语句时，在不知道第一条语句的情况下，我们无法判断是谁捐钱给穷人。因此，我们应该将 He 与 Ravi 联系起来才能得到完整的句子意义。这就是我们的大脑在进行的指代消解（reference resolution）过程。

如果我们仔细观察前面的例子，首先出现主语，然后出现代词。所以流动（flow）的方向是从左到右。基于这种流动，我们可以称这些类型的句子为回指（anaphora）。

让我们再举一个例子：

He was already on his way to airport. Realized Ravi

这是另一种类型的例子，表达的方向是逆序（首先是代词，然后是名词）。这里，同样是 He 与 Ravi 相关联。这种类型的句子称为预指（Cataphora）。

最早的指代消解算法可以追溯到 1970 年，霍布斯（Hobbs）发表了一篇相关论文。这篇论文的网络版本可以通过以下网址获得：https:/ /www. isi. edu/ ~hobbs/ pronoun- papers. html.

在本节中，我们将利用刚才所学的内容来编写一个非常简单的指代消解算法。

8.6.1　准备工作

首先，你需要安装 Python 工具包、NLTK 库和 gender 数据集。你可以利用 nltk.download() 下载语料。

8.6.2　如何实现

1. 打开 Atom 编辑器（或者你常用的程序编辑器）。
2. 创建一个名为 Anaphora.py 的新文件。
3. 输入以下源代码：

```
Anaphora.py
1  import nltk
2  from nltk.chunk import tree2conlltags
3  from nltk.corpus import names
4  import random
5
6  class AnaphoraExample:
7      def __init__(self):
8          males = [(name, 'male') for name in names.words('male.txt')]
9          females = [(name, 'female') for name in names.words('female.txt')]
10         combined = males + females
11         random.shuffle(combined)
12         training = [(self.feature(name), gender) for (name, gender) in combined]
13         self._classifier = nltk.NaiveBayesClassifier.train(training)
14
15     def feature(self, word):
16         return {'last(1)' : word[-1]}
17
18     def gender(self, word):
19         return self._classifier.classify(self.feature(word))
```

```
20
21      def learnAnaphora(self):
22          sentences = [
23              "John is a man. He walks",
24              "John and Mary are married. They have two kids",
25              "In order for Ravi to be successful, he should follow John",
26              "John met Mary in Barista. She asked him to order a Pizza"
27          ]
28
29          for sent in sentences:
30              chunks = nltk.ne_chunk(nltk.pos_tag(nltk.word_tokenize(sent)),
                binary=False)
31              stack = []
32              print(sent)
33              items = tree2conlltags(chunks)
34              for item in items:
35                  if item[1] == 'NNP' and (item[2] == 'B-PERSON' or item[2] ==
                    'O'):
36                      stack.append((item[0], self.gender(item[0])))
37                  elif item[1] == 'CC':
38                      stack.append(item[0])
39                  elif item[1] == 'PRP':
40                      stack.append(item[0])
41              print("\t {}".format(stack))
42
43  anaphora = AnaphoraExample()
44  anaphora.learnAnaphora()
45
```

4. 保存文件。
5. 使用 Python 编译器运行程序。
6. 你将看到如下输出：

```
ch8 — -bash — 104×23
nltk $ python Anaphora.py
John is a man. He walks
        [('John', 'male'), 'He']
John and Mary are married. They have two kids
        [('John', 'male'), 'and', ('Mary', 'female'), 'They']
In order for Ravi to be successful, he should follow John
        [('Ravi', 'female'), 'he', ('John', 'male')]
John met Mary in Barista. She asked him to order a Pizza
        [('John', 'male'), ('Mary', 'female'), 'She', 'him']
nltk $
```

8.6.3 工作原理

我们来理解这个简单的指代消解算法的工作原理：

```
import nltk
from nltk.chunk import tree2conlltags
from nltk.corpus import names
import random
```

以上四条代码导入了程序中必需的模块和函数。我们定义一个名为 anaphoraexample 的类：

```
class AnaphoraExample:
```

为这个类定义一个不接受任何参数的构造函数：

```
def __init__(self):
```

以下两条代码从 nltk.names 语料库下载所有男性和女性的名字，将它们标记为 male/female，并存储在名为 males/females 的两个列表中：

```
males = [(name, 'male') for name in names.words('male.txt')]
females = [(name, 'female') for name in names.words('female.txt')]
```

以下这条代码由 males 列表和 females 列表创建了一个唯一列表。而 random.shuffle() 函数使得列表中的数据随机分布：

```
combined = males + females
random.shuffle(combined)
```

这条代码在 gender 数据上调用 feature() 函数，并将所有的名字存储在名为 training 的变量中：

```
training = [(self.feature(name), gender) for (name, gender) in
combined]
```

我们利用保存在 training 变量中的男性和女性特征，创建了一个朴素贝叶斯分类器（NaiveBayesClassifier）对象，名为 _classifier：

```
self._classifier = nltk.NaiveBayesClassifier.train(training)
```

以下函数定义了一个最简单的特征，即仅通过给定名字的最后一个字母来判断是男性还是女性的名字：

```
def feature(self, word):
    return {'last(1)' : word[-1]}
```

以下函数将一个单词作为参数，并尝试使用我们构建的分类器来检测人物的性别：

```
def gender(self, word):
    return self._classifier.classify(self.feature(word))
```

以下是在例句中检测指代的主要函数：

```
def learnAnaphora(self):
```

以下是四个以指代形式表达的混合复杂句例子：

```
sentences = [
    "John is a man. He walks",
    "John and Mary are married. They have two kids",
    "In order for Ravi to be successful, he should follow John",
    "John met Mary in Barista. She asked him to order a Pizza"
]
```

以下代码遍历所有句子，每一次遍历都将当前句子赋值给名为 sent 的局部变量：

```
for sent in sentences:
```

以下代码实现句子分词、词性标注、组块（实体）抽取，并返回组块树结果给名为

chunks 的变量：

```
chunks = nltk.ne_chunk(nltk.pos_tag(nltk.word_tokenize(sent)),
binary=False)
```

以下变量存储所有用于解决指代问题的名字和代词：

```
stack = []
```

以下代码在用户屏幕上显示正在处理的当前语句：

```
print(sent)
```

以下代码将组块树（chunks）展平成一个列表，并以 IOB 格式表示：

```
items = tree2conlltags(chunks)
```

遍历分块后以 IOB 格式标注的所有句子（有 3 个元素的元组）：

```
for item in items:
```

如果单词的词性是 NNP，并且该单词的 IOB 格式是 B-PERSON 或 O，那么我们将标记这个单词为人名：

```
if item[1] == 'NNP' and (item[2] == 'B-PERSON' or item[2]
== 'O'):
stack.append((item[0], self.gender(item[0])))
```

如果单词的词性是 CC，我们将它加入到 stack 变量中：

```
elif item[1] == 'CC':
    stack.append(item[0])
```

如果单词的词性是 PRP，我们也将它加入到 stack 变量中：

```
elif item[1] == 'PRP':
    stack.append(item[0])
```

最后，在屏幕上打印出 stack 变量：

```
print("\t {}".format(stack))
```

以下代码从 anaphoraexample() 中创建了一个名为 anaphora 的对象，并在 anaphora 对象上调用 learnanaphora() 函数。当执行完成这个函数，我们将看到每条句子的单词列表：

```
anaphora = AnaphoraExample()
anaphora.learnAnaphora()
```

8.7　词义消歧

在前面章节中，我们已经学习了如何进行单词的词性标注和命名实体识别。在英语中，一个单词有可能同时有名词词性和动词词性，判断一个单词的词性对计算机程序来说是相当困难的。

我们列举一些例子来理解词义的概念：

例句	说明
She is my date（她和我约会。）	在这里，date 的词义并不是日历中的日期，而是表示一种人际关系——约会。
You have taken too many leaves to skip cleaning leaves in the garden（为了不用清理花园里的树叶，你已经多次以休息为由逃避。）	在这里，leaves 有多重词义： • 第一次出现的 leave 代表休息、告退。 • 第二次出现的 leave 指树上的树叶。

如上例所示，句子中可能存在单词多重词义的组合情况。

对于词义判断，我们目前面临的挑战之一就是找到一种适当的命名法则来描述这些词义。现在有很多介绍单词用法以及各种搭配情况的英文词典，而 WordNet 是目前为止结构化最好且被广泛认可的词义用法资源库。

在本节中，我们将展示一些从 WordNet 库中选取的范例，并使用内置的 NLTK 库来确定这些词的词义。

Lesk 算法是用来解决词义判断问题的最古老的算法。但是，你将会发现，在某些情况下 lesk 算法得到的结果也不那么准确。

8.7.1　准备工作

安装 Python 和 NLTK 库。

8.7.2　如何实现

1. 打开 Atom 编辑器（或者你常用的程序编辑器）。
2. 创建一个新文件，命名为 WordSense.py。
3. 输入以下源代码：

```
WordSense.py
1  import nltk
2
3  def understandWordSenseExamples():
4      words = ['wind', 'date', 'left']
5      print("-- examples --")
6      for word in words:
7          syns = nltk.corpus.wordnet.synsets(word)
8          for syn in syns[:2]:
9              for example in syn.examples()[:2]:
10                 print("{} -> {} -> {}".format(word, syn.name(), example))
11
12
13 def understandBuiltinWSD():
14     print("-- built-in wsd --")
15     maps = [
16         ('Is it the fish net that you are using to catch fish ?', 'fish', 'n'),
17         ('Please dont point your finger at others.', 'point', 'n'),
```

```
18        ('I went to the river bank to see the sun rise', 'bank', 'n'),
19    ]
20    for m in maps:
21        print("Sense '{}' for '{}' -> '{}'".format(m[0], m[1],
 .        nltk.wsd.lesk(m[0], m[1], m[2])))
22
23 if __name__ == '__main__':
24     understandWordSenseExamples()
25     understandBuiltinWSD()
26
```

4. 保存文件。
5. 使用 Python 编译器运行程序。
6. 你将看到如下输出：

```
ch8 — -bash — 104×23
nltk $ python WordSense.py
-- examples --
wind -> wind.n.01 -> trees bent under the fierce winds
wind -> wind.n.01 -> when there is no wind, row
wind -> wind.n.02 -> the winds of change
date -> date.n.01 -> what is the date today?
date -> date.n.02 -> his date never stopped talking
left -> left.n.01 -> she stood on the left
-- built-in wsd --
Sense 'Is it the fish net that you are using to catch fish ?' for 'fish' -> 'Synset('pisces.n.02')'
Sense 'Please dont point your finger at others.' for 'point' -> 'Synset('point.n.25')'
Sense 'I went to the river bank to see the sun rise' for 'bank' -> 'Synset('savings_bank.n.02')'
nltk $
```

8.7.3　工作原理

下面我们来分析程序是如何实现的。首先，下面这行代码用于导入 NLTK 库：

```
import nltk
```

然后，定义一个名为 understandWordSenseExamples() 的函数，该函数能够通过查阅 WordNet 语料库来输出我们感兴趣的单词的可能词义：

```
def understandWordSenseExamples():
```

以下三个单词都有多重词义，它们被保存到一个名为 words 的列表中：

```
 words = ['wind', 'date', 'left']

print("-- examples --")
 for word in words:
     syns = nltk.corpus.wordnet.synsets(word)
     for syn in syns[:2]:
         for example in syn.examples()[:2]:
             print("{} -> {} -> {}".format(word, syn.name(), example))
```

上述代码执行如下操作：

- 遍历列表中所有的单词，并将当前单词保存到 word 变量中；
- 调用 wordnet 模块中的 synsets() 函数，并将运算结果保存到 syns 变量中；
- 从列表中获取并遍历前 3 个同义词集（synsents），将当前同义词集保存到 syn 变量中；
- 调用 syn 对象中的 examples() 函数，并选取前两个例子进行遍历，每次遍历时的当前值保存到 example 变量中；
- 在屏幕上输出单词、同义词集的名称以及例句。

为了观察 NLTK 库内置的 lesk 算法对例句的处理性能，我们定义一个新的函数 understand BuiltinWSD()：

```
def understandBuiltinWSD():
```

定义一个新的元组列表变量 maps：

```
    print("-- built-in wsd --")
    maps = [
        ('Is it the fish net that you are using to catch fish ?', 'fish',
'n'),
        ('Please dont point your finger at others.', 'point', 'n'),
        ('I went to the river bank to see the sun rise', 'bank', 'n'),
    ]
```

每个元组包含三个元素：

- 待分析的句子
- 句子中需要判断词义的单词
- 单词的词性信息

下列两行代码遍历 maps 变量，并将遍历的当前元组赋值给 m 变量，然后调用 nltk.wsd.lesk() 函数，最后在屏幕上输出格式化结果：

```
    for m in maps:
        print("Sense '{}' for '{}' -> '{}'".format(m[0], m[1],
nltk.wsd.lesk(m[0], m[1], m[2])))
```

当运行主程序时，调用如下两个函数，并在屏幕上输出结果。

```
if __name__ == '__main__':
    understandWordSenseExamples()
    understandBuiltinWSD()
```

8.8　情感分析

信息反馈是人际关系理解中最有效的措施之一。人们很擅长在语言沟通交流的过程中理解反馈信息，而这些理解分析过程是无意识的。为了编写出能够衡量人类情感的计算机程序，我们需要较好地理解这些情感在自然语言中的表达方式。

我们来看这些例子：

例句	描述
I am very happy（我很开心）	表示一种开心的情感
She is so :(（她很难过）	这是一种表达悲伤难过的图标

在自然语言的书面交流中，随着文本、图标和表情（emojis）使用频率的增加，计算机程序越来越难理解句子的情感含义。

下面，我们编写一个程序来了解如何利用 NLTK 库提供的工具构建我们自己的算法。

8.8.1 准备工作

安装 Python 和 NLTK 库。

8.8.2 如何实现

1. 打开 Atom 编辑器（或者你常用的程序编辑器）。
2. 创建一个新文件，命名为 Sentiment.py。
3. 输入以下源代码：

```
Sentiment.py
1  import nltk
2  import nltk.sentiment.sentiment_analyzer
3
4  def wordBasedSentiment():
5      positive_words = ['love', 'hope', 'joy']
6      text = 'Rainfall this year brings lot of hope and joy to Farmers.'.split()
7      analysis = nltk.sentiment.util.extract_unigram_feats(text, positive_words)
8      print(' -- single word sentiment --')
9      print(analysis)
10
11 def multiWordBasedSentiment():
12     word_sets = [('heavy', 'rains'), ('flood', 'bengaluru')]
13     text = 'heavy rains cause flash flooding in bengaluru'.split()
14     analysis = nltk.sentiment.util.extract_bigram_feats(text, word_sets)
15     print(' -- multi word sentiment --')
16     print(analysis)
17
18 def markNegativity():
19     text = 'Rainfall last year did not bring joy to Farmers'.split()
20     negation = nltk.sentiment.util.mark_negation(text)
21     print(' -- negativity --')
22     print(negation)
23
24 if __name__ == '__main__':
25     wordBasedSentiment()
26     multiWordBasedSentiment()
27     markNegativity()
28
```

4. 保存文件。

5. 使用 Python 编译器运行程序。

6. 你将看到如下输出：

```
ch8 — -bash — 104×23
nltk $ atom Sentiment.py
nltk $ python Sentiment.py
 -- single word sentiment --
{'contains(love)': False, 'contains(hope)': True, 'contains(joy)': True}
 -- multi word sentiment --
{'contains(heavy - rains)': True, 'contains(flood - bengaluru)': False}
 -- negativity --
['Rainfall', 'last', 'year', 'did', 'not', 'bring_NEG', 'joy_NEG', 'to_NEG', 'Farmers_NEG']
nltk $
```

8.8.3　工作原理

现在，我们来看看情感分析程序是如何工作的。以下代码分别导入了 NLTK 模块和 sentiment_analyzer 模块：

```
import nltk
import nltk.sentiment.sentiment_analyzer
```

定义一个新的函数 wordBasedSentiment()，用于学习如何基于已知的词语进行情感分析，这些词语对我们来说具有重要意义：

```
def wordBasedSentiment():
```

现在我们定义一个包含 3 个单词的列表，这些单词特别表达了快乐的某种形式，存储在 positive_words 变量中：

```
positive_words = ['love', 'hope', 'joy']
```

这是待分析的样例文本，该文本存储在名为 text 的变量中：

```
text = 'Rainfall this year brings lot of hope and joy to
Farmers.'.split()
```

我们在 text 变量上调用 extract_unigram_feats() 函数，并以我们已经定义的单词为参数，输出结果是一个字典，用于指示给定单词是否出现在文本中：

```
analysis = nltk.sentiment.util.extract_unigram_feats(text,
positive_words)
```

以下代码在用户的屏幕上显示字典：

```
print(' -- single word sentiment --')
print(analysis)
```

以下代码定义了一个新函数，用于判断一个句子中是否存在某些单词对：

```
def multiWordBasedSentiment():
```

该代码定义了一个双词元组（two-word tuples）的列表。我们需要在一个句子中查找这些单词对是否在一个句子中同时出现：

```
word_sets = [('heavy', 'rains'), ('flood', 'bengaluru')]
```

以下是待处理和待查找特征的句子：

```
text = 'heavy rains cause flash flooding in bengaluru'.split()
```

在输入句子上调用 extract_bigram_feats() 函数，查找其中是否包含 word_sets 变量中的单词集，返回结果是一个字典，表明这些单词对是否存在于句子中：

```
analysis = nltk.sentiment.util.extract_bigram_feats(text, word_sets)
```

以下代码在屏幕上打印出字典：

```
print(' -- multi word sentiment --')
print(analysis)
```

现在我们定义一个新函数 markNegativity()，找出句子的负向性（negativity）：

```
def markNegativity():
```

以下是需要进行负向性分析的句子，被存储在 text 变量中：

```
text = 'Rainfall last year did not bring joy to Farmers'.split()
```

在 text 上调用 mark_negation() 函数，返回句子中所有单词的列表，并在属于负向性含义的单词后附加一个特殊后缀"_NEG"，结果存储在 negation 变量中：

```
negation = nltk.sentiment.util.mark_negation(text)
```

该代码在屏幕上打印出 negation 列表：

```
print(' -- negativity --')
print(negation)
```

当程序运行时，调用以下三个函数，我们将看到这些函数的执行顺序（自顶向下）：

```
if __name__ == '__main__':
    wordBasedSentiment()
    multiWordBasedSentiment()
    markNegativity()
```

8.9 高阶情感分析

越来越多的企业为了增加它们的目标用户群，开通了网上交易，这样用户可以通过各种渠道留下反馈信息。因此对企业来说，了解用户对他们经营业务的情感反馈变得越来越重要。

在本节中，我们将基于前一节所学内容来编写情感分析程序，探索内置的 vader 情感分析算法，并了解复杂句子的情感分析过程。

8.9.1　准备工作

安装 Python 和 NLTK 库。

8.9.2　如何实现

1. 打开 Atom 编辑器（或者你常用的程序编辑器）。
2. 创建一个名为 AdvSentiment.py 的新文件。
3. 输入以下源代码：

```
AdvSentiment.py
1   import nltk
2   import nltk.sentiment.util
3   import nltk.sentiment.sentiment_analyzer
4   from nltk.sentiment.vader import SentimentIntensityAnalyzer
5
6   def mySentimentAnalyzer():
7       def score_feedback(text):
8           positive_words = ['love', 'genuine', 'liked']
9           if '_NEG' in ' '.join(nltk.sentiment.util.mark_negation(text.split())):
10              score = -1
11          else:
12              analysis = nltk.sentiment.util.extract_unigram_feats(text.split(),
.               positive_words)
13              if True in analysis.values():
14                  score = 1
15              else:
16                  score = 0
17          return score
18
19      feedback = """I love the items in this shop, very genuine and quality is
.       well maintained.
20      I have visited this shop and had samosa, my friends liked it very much.
21      ok average food in this shop.
22      Fridays are very busy in this shop, do not place orders during this day."""
23      print(' -- custom scorer --')
24      for text in feedback.split("\n"):
25          print("score = {} for >> {}".format(score_feedback(text), text))
26
27
28  def advancedSentimentAnalyzer():
29      sentences = [
30          ':)',
31          ':(',
```

```
32            'She is so :(',
33            'I love the way cricket is played by the champions',
34            'She neither likes coffee nor tea',
35        ]
36        senti = SentimentIntensityAnalyzer()
37        print(' -- built-in intensity analyser --')
38        for sentence in sentences:
39            print('[{}]'.format(sentence), end=' --> ')
40            kvp = senti.polarity_scores(sentence)
41            for k in kvp:
42                print('{} = {}, '.format(k, kvp[k]), end='')
43            print()
44
45    if __name__ == '__main__':
46        advancedSentimentAnalyzer()
47        mySentimentAnalyzer()
48
```

4. 保存文件。
5. 使用 Python 编译器运行程序。
6. 你将看到如下输出：

```
ch8 — -bash — 104×23
nltk $ python AdvSentiment.py
 -- built-in intensity analyser --
[:)] --> neg = 0.0, neu = 0.0, pos = 1.0, compound = 0.4588,
[:(] --> neg = 1.0, neu = 0.0, pos = 0.0, compound = -0.4404,
[She is so :(] --> neg = 0.555, neu = 0.445, pos = 0.0, compound = -0.5777,
[I love the way cricket is played by the champions] --> neg = 0.0, neu = 0.375, pos = 0.625, compound =
0.875,
[She neither likes coffee nor tea] --> neg = 0.318, neu = 0.682, pos = 0.0, compound = -0.3252,
 -- custom scorer --
score = 1 for >> I love the items in this shop, very genuine and quality is well maintained.
score = 1 for >>     I have visited this shop and had samosa, my friends liked it very much.
score = 0 for >>     ok average food in this shop.
score = -1 for >>     Fridays are very busy in this shop, do not place orders during this day.
nltk $
```

8.9.3　工作原理

现在，我们来分析情感分析程序是如何工作的。以下四条代码导入程序必需的模块：

```
import nltk
import nltk.sentiment.util
import nltk.sentiment.sentiment_analyzer
from nltk.sentiment.vader import SentimentIntensityAnalyzer
```

定义一个新函数 mySentimentAnalyzer()：

```
def mySentimentAnalyzer():
```

下面这条代码定义了一个新的子函数 score_feedback()，它将一个句子作为输入，并返回句子得分，其中 -1 为负，0 为中性，1 为正：

```
def score_feedback(text):
```

为了实验需要，我们仅定义如下三个单词用于情感判断。在现实情况中，我们可能会从较大的字典语料库中选取单词：

```
positive_words = ['love', 'genuine', 'liked']
```

下面的代码对输入的句子进行分词，之后将单词列表提供给 mark_negation() 函数以识别句子中是否存在负向性（negativity）。将 mark_negation() 函数的返回结果拼接成字符串，查看是否存在 _NEG 后缀，如果存在则将分数设置为 -1：

```
        if '_NEG' in '
'.join(nltk.sentiment.util.mark_negation(text.split())):
            score = -1
```

这里，我们调用 extract_unigram_feats() 来查找输入文本中的正向词（positive_words），并将返回的字典存储到名为 analysis 的变量中：

```
        else:
            analysis =
nltk.sentiment.util.extract_unigram_feats(text.split(), positive_words)
```

如果输入文本中存在正向词，则分数被置为 1：

```
if True in analysis.values():
    score = 1
else:
    score = 0
```

最后，score_feedback() 函数返回计算得分：

```
return score
```

以下是四条用户反馈，我们利用算法进行计算并输出分数：

```
    feedback = """I love the items in this shop, very genuine and quality
is well maintained.
    I have visited this shop and had samosa, my friends liked it very much.
    ok average food in this shop.
    Fridays are very busy in this shop, do not place orders during this
day."""
```

以下代码通过使用换行符（\n）将 feedback 变量分割成句子，并对每个句子调用 score_feedback() 函数：

```
print(' -- custom scorer --')
for text in feedback.split("\n"):
    print("score = {} for >> {}".format(score_feedback(text), text))
```

输出结果为句子和对应分数。以下代码定义了 advancedSentimentAnalyzer() 函数，便于我们了解 NLTK 内置的情感分析工具：

```
def advancedSentimentAnalyzer():
```

接下来定义五个待分析句子。不难发现，我们还使用了表情图标来测试算法是如何工作的：

```
sentences = [
    ':)',
    ':(',
    'She is so :(',
    'I love the way cricket is played by the champions',
    'She neither likes coffee nor tea',
]
```

以下代码为 SentimentIntensityAnalyzer() 创建了一个新的对象，并将该对象存储在变量 senti 中：

```
  senti = SentimentIntensityAnalyzer()

print(' -- built-in intensity analyser --')
  for sentence in sentences:
      print('[{}]'.format(sentence), end=' --> ')
      kvp = senti.polarity_scores(sentence)
      for k in kvp:
          print('{} = {}, '.format(k, kvp[k]), end='')
      print()
```

上述代码所做工作如下：

- 遍历所有句子，并将当前句子存储到 sentence 变量中；
- 在屏幕上显示当前处理的句子；
- 调用 polarity_scores() 函数处理句子，并将结果存储在一个名为 kvp 的变量中；
- 遍历 kvp 字典并打印出键（负向性、中性、正向性或复合类型）以及基于相应类型的计算得分。

运行当前程序时，调用以下两个函数并在屏幕上输出结果：

```
if __name__ == '__main__':
    advancedSentimentAnalyzer()
    mySentimentAnalyzer()
```

8.10 创建一个对话助手或聊天机器人

对我们来说，对话助手或聊天机器人并不算新颖。值得一提的是其中最重要的一种对话助手 ELIZA，它创建于 20 世纪 60 年代初。

为了成功构建一个对话引擎，我们需要注意以下几点：

- 了解目标用户
- 理解用于沟通的自然语言
- 了解用户的意图
- 应答用户，并给出进一步的线索

NLTK 有一个 nltk.chat 模块，它提供一个通用的框架，可以简化对话引擎的构建。

NLTK 中可用的引擎如下表所示：

引擎	模块
Eliza	nltk.chat.eliza Python 模块
Iesha	nltk.chat.iesha Python 模块
Rude	nltk.chat.rudep Python 模块
Suntsu	nltk.chat. suntsu 模块
Zen	nltk.chat.zen 模块

为了与这些引擎进行交互，我们可以在 Python 程序中加载这些模块，并调用 demo() 函数。

本节向我们展示了如何使用内置引擎，并使用 nltk.chat 模块提供的框架编写我们自己的简单对话引擎。

8.10.1　准备工作

安装 Python 和 NLTK 库，掌握正则表达式将有所帮助。

8.10.2　如何实现

1. 打开 Atom 编辑器（或者你常用的程序编辑器）。
2. 创建一个名为 Conversational.py 的新文件。
3. 输入以下源代码：

```
Conversational.py
1  import nltk
2
3  def builtinEngines(whichOne):
4      if whichOne == 'eliza':
5          nltk.chat.eliza.demo()
6      elif whichOne == 'iesha':
7          nltk.chat.iesha.demo()
8      elif whichOne == 'rude':
9          nltk.chat.rude.demo()
10     elif whichOne == 'suntsu':
```

```
        nltk.chat.suntsu.demo()
    elif whichOne == 'zen':
        nltk.chat.zen.demo()
    else:
        print("unknown built-in chat engine {}".format(whichOne))

def myEngine():
    chatpairs = (
        (r"(.*?)Stock price(.*)",
            ("Today stock price is 100",
            "I am unable to find out the stock price.")),
        (r"(.*?)not well(.*)",
            ("Oh, take care. May be you should visit a doctor",
            "Did you take some medicine ?")),
        (r"(.*?)raining(.*)",
            ("Its monsoon season, what more do you expect ?",
            "Yes, its good for farmers")),
        (r"How(.*?)health(.*)",
            ("I am always healthy.",
            "I am a program, super healthy!")),
        (r".*",
            ("I am good. How are you today ?",
            "What brings you here ?"))
    )
    def chat():
        print("!"*80)
        print(" >> my Engine << ")
        print("Talk to the program using normal english")
        print("="*80)
        print("Enter 'quit' when done")
        chatbot = nltk.chat.util.Chat(chatpairs, nltk.chat.util.reflections)
        chatbot.converse()

    chat()

if __name__ == '__main__':
    for engine in ['eliza', 'iesha', 'rude', 'suntsu', 'zen']:
        print("=== demo of {} ===".format(engine))
        builtinEngines(engine)
        print()
    myEngine()
```

4. 保存文件。

5. 使用 Python 编译器运行程序。

6. 你将看到如下输出：

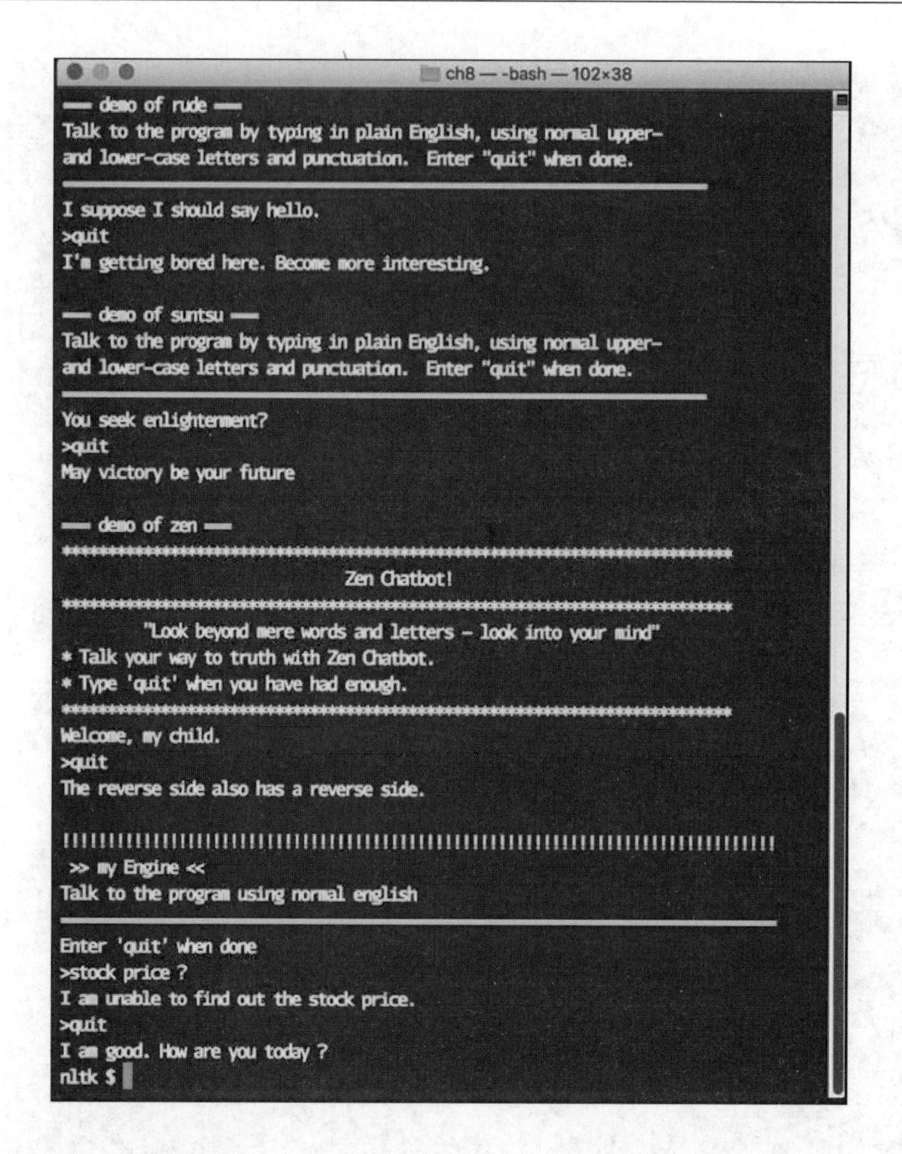

8.10.3　工作原理

现在，我们来分析程序是如何工作的。以下代码将 NLTK 库导入到当前程序中：

```
import nltk
```

以下代码定义了一个名为 builtinEngines 的新函数，它的输入是一个字符串参数 whichOne：

```
def builtinEngines(whichOne):
```

以下这些 if、elif、else 代码是典型的分支代码，根据 whichOne 变量参数来决定调用哪种对话引擎的 demo() 函数。当用户输入一个未知的引擎名时，程序会向用户返回没有该引擎的消息：

```
if whichOne == 'eliza':
    nltk.chat.eliza.demo()
elif whichOne == 'iesha':
    nltk.chat.iesha.demo()
elif whichOne == 'rude':
    nltk.chat.rude.demo()
elif whichOne == 'suntsu':
    nltk.chat.suntsu.demo()
elif whichOne == 'zen':
    nltk.chat.zen.demo()
else:
    print("unknown built-in chat engine {}".format(whichOne))
```

在实践中需要考虑所有已知和未知的情况，这使程序在处理未知情况时更加具有鲁棒性。

以下代码定义了一个名为 myEngine() 的新函数，该函数不带任何参数：

```
def myEngine():
```

以下代码定义了一个嵌套的元组数据结构，并将其赋值给 chatpairs 变量：

```
chatpairs = (
    (r"(.*?)Stock price(.*)",
        ("Today stock price is 100",
        "I am unable to find out the stock price.")),
    (r"(.*?)not well(.*)",
        ("Oh, take care. May be you should visit a doctor",
        "Did you take some medicine ?")),
    (r"(.*?)raining(.*)",
        ("Its monsoon season, what more do you expect ?",
        "Yes, its good for farmers")),
    (r"How(.*?)health(.*)",
        ("I am always healthy.",
        "I am a program, super healthy!")),
    (r".*",
        ("I am good. How are you today ?",
        "What brings you here ?"))
)
```

接下来我们详细介绍该数据结构：

- 我们定义了一个元组的元组
- 每个子元组包含两个元素：
    - 第一个成员是一个正则表达式（以正则表达式格式存储的用户问题）
    - 第二个成员是另一组元组（用户问题的答案）

我们在 myEngine() 函数中定义了一个名为 chat() 的子函数。chat() 函数在屏幕上向用户显示一些信息，并调用 NLTK 内置的 nltk.chat.util.Chat() 类，输入参数依次为变量 chatpairs 和 nltk.chat.util.reflections。最后，我们在使用 chat() 类创建的对象上调用函数 chatbot.converse()：

```
    def chat():
        print("!"*80)
        print(" >> my Engine << ")
        print("Talk to the program using normal english")
        print("="*80)
        print("Enter 'quit' when done")
        chatbot = nltk.chat.util.Chat(chatpairs,
nltk.chat.util.reflections)
        chatbot.converse()
```

以下代码调用了 chat() 函数，该函数在屏幕上打印出提示并接收用户请求，并根据我们之前建立的正则表达式来做出反应：

```
chat()
```

当作为一个独立程序调用（不使用导入）程序时，则添加以下代码；

```
if __name__ == '__main__':
    for engine in ['eliza', 'iesha', 'rude', 'suntsu', 'zen']:
        print("=== demo of {} ===".format(engine))
        builtinEngines(engine)
        print()
    myEngine()
```

上述代码所做工作有两部分：

- 依次逐个调用内置引擎（方便我们进行体验）；
- 运行五个内置引擎后，则通过调用 myEngine() 函数来运行我们的用户引擎。

CHAPTER 9

第9章 深度学习在自然语言处理中的应用

9.1 引言

近年来，深度学习在文本、语音和图像领域取得了突出成果，这些成果主要用于人工智能领域的应用开发上。不但如此，深度学习模型在其他应用领域中也做出了突出的贡献。在本章中，我们将介绍深度学习在**自然语言处理**及文本处理领域中的多种应用。

卷积神经网络（Convolutional Neural Network，CNN）和**循环神经网络**（Recurrent Neural Network，RNN）是深度学习中的核心主题，你将在深度学习领域中不断接触到这些网络。

9.1.1 卷积神经网络

CNN 主要用于图像处理，常用来解决图像分类问题。CNN 的工作原理如下图所示，利用 3×3 大小的滤波器对大小为 5×5 的原始矩阵进行卷积，输出大小为 3×3 的矩阵。滤波器可以以步长为 1 或大于 1 的任意值进行水平移动。对于输出矩阵的单元格（1，1），其值为 3，这是由原始矩阵和滤波器做乘积运算得到的。运用同样的方式，滤波器将游走在原始的 5×5 矩阵上以创建 3×3 的卷积特征，也被称为特征图或激活映射（activation map）：

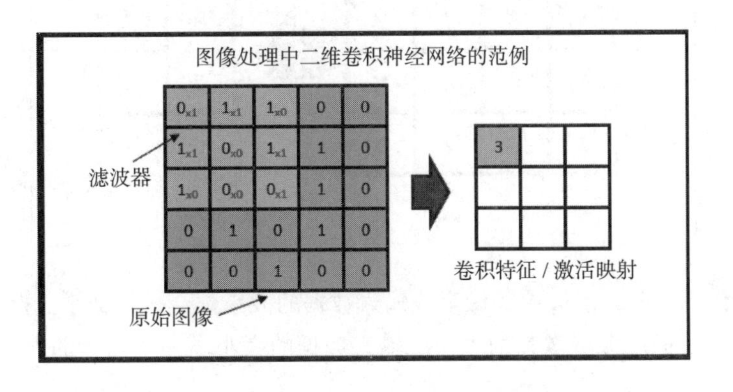

使用卷积的优势：

- 全连接层代替固定尺寸，减少了神经元的数量，也因此降低了对机器计算能力的需求。
- 通过使用小尺寸的滤波器对原始矩阵进行卷积，避免了每个像素都连接到下一层。因此，对需要将输入图像汇总到下一层的任务来说，这是一种更好的方法。
- 反向传播过程中，只需根据反向传播误差更新滤波器的权重，就可以使效率得到提升。

CNN 在任意维数的空间或时间分布式数组上都可以进行映射。因此，CNN 适合应用于时间序列、图像或视频领域。CNN 具有以下特征：

- 平移不变性（空间转换时，神经元权重固定不变）
- 局部连通性（仅在空间局部区域之间存在神经连接）
- 空间分辨率的可选渐进式下降（随着特征的数量逐渐增加）

完成卷积操作之后，在保留最重要特征的基础上仍然要对卷积特征 / 激活映射进行缩减。因为这样的操作不仅减少了点的数量，而且提高了计算效率。池化（pooling）操作通常用于减少不必要的表示。有关池化操作的简要介绍如下：

- **池化**：池化操作使激活表示（activation representation）（由滤波器对输入矩阵和权值进行卷积而得到）的维数减小，更易于管理。池化与激活映射操作相互独立。池化只作用于特征层的宽度和广度，对其深度不做处理。下面的图展示了大小为 2 × 2 的最大池化操作方法，我们可以看到下面的这个 4 × 4 的原始矩阵减小了一半幅度。具体来说，在第一个 2 × 2 的单元中，值分别为 2、4、5、8，池化操作将其中最大的值 8 抽取出来，按照同样的方法对原始矩阵的其他部分进行池化，得到右面的结果矩阵：

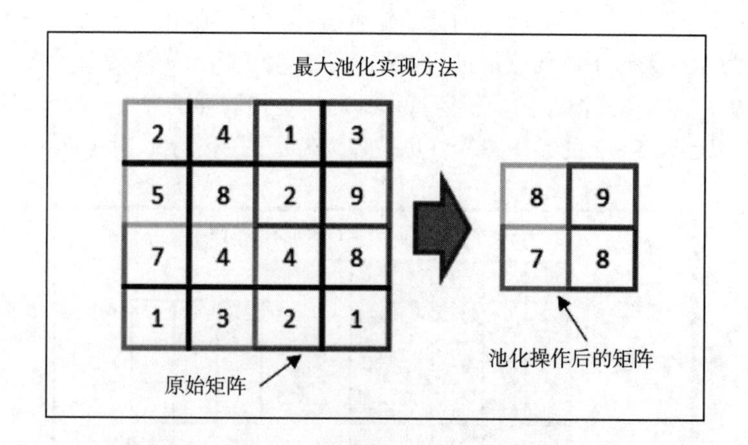

在卷积操作下，各个阶段中的像素或输入数据的大小都会自然减小。但在某些情况下，我们想要在操作后仍然保持像素的大小或输入数据的大小。一个巧妙的方法就是在边界相应地填充 0。

- **填充**：下面图表的宽度和广度在卷积操作下会连续收缩，这在深度网络中是不可取的。填充可以保持图片大小不发生改变或者使整个网络的大小处于可控状态。

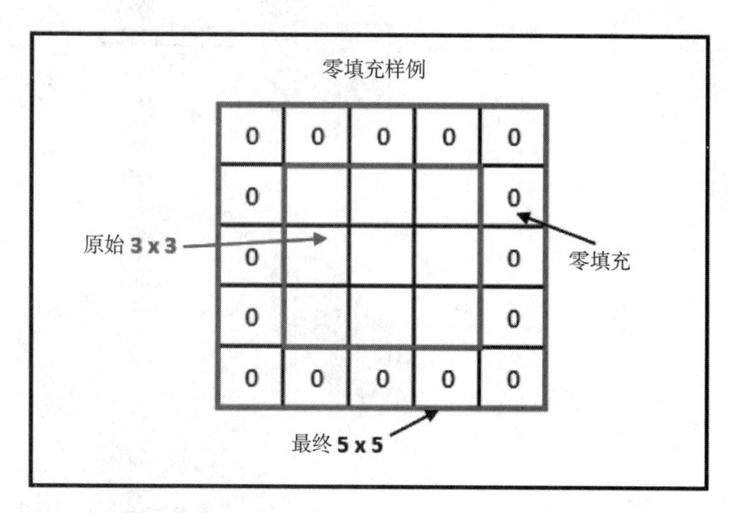

基于给定的输入宽度、滤波器大小、填充和步长，计算特征图大小的公式如下所示，同时这个公式给出了所需计算能力大小的度量方法。

- **特征图大小的计算**：在下面的公式中，从卷积层得到的特征图大小是：

$$\text{Activation Map Size}=\left(\frac{W-F+2P}{S}\right)+1$$

其中，W 是原始图像的宽度，F 是滤波器的大小，P 是填充大小（1 为单层填充，2 为双层填充，依此类推），S 是步长。

例如，对于一幅大小为 $224\times224\times3$（3 表示红色、绿色和蓝色三个颜色通道）的输入图像，滤波器大小为 11×11，滤波器数为 96。步长为 4，而且没有填充。那么由已知条件怎么确定特征图大小呢？

$$\text{Activation Map Size}=\left(\frac{W-F+2P}{S}\right)+1$$

$$\text{Activation Map Size}=\left(\frac{224-11+2*0}{4}\right)+1=54.25\sim55$$

特征图的维数为 $55\times55\times96$。利用前面的公式，只能计算出宽度和广度，深度取决于所用滤波器的个数。事实上，这是使用 AlexNet 类型进行卷积操作的结果，现在我们对其进行描述。

- **AlexNet 参加了 2012 年的 ImageNet 竞赛**：下图描述了 AlexNet，它在 2012 年的 ImageNet 竞赛中赢得胜利。与其他竞争者相比，AlexNet 取得了更高的精确度。

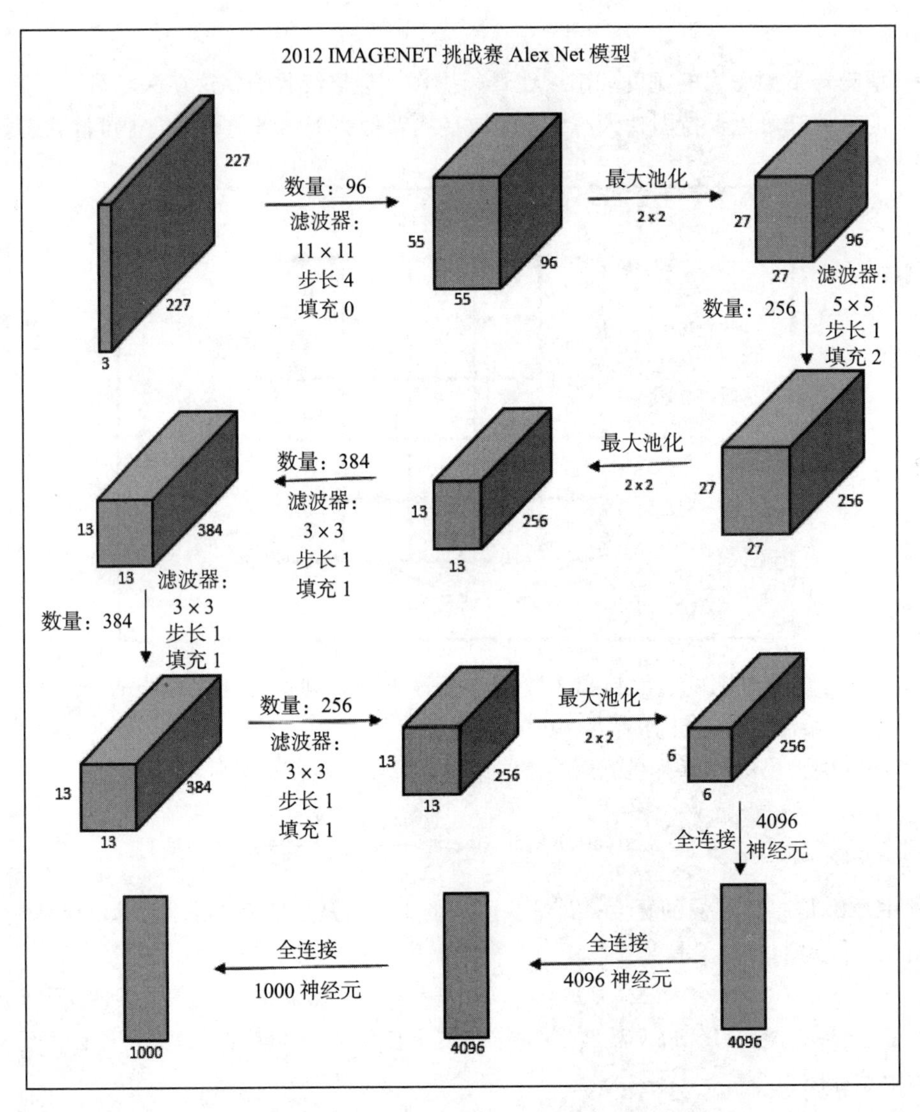

AlexNet 巧妙地运用了卷积、池化、填充等技巧，并最终与全连接层相连。

CNNs 的应用

CNN 有各种不同的应用领域，下面进行举例介绍：

- **图像分类**：与其他方法相比，CNN 在大规模图像数据上的实验取得了更高的准确率。在图像分类的初始阶段，通过叠加多层 CNN，抽取到足够多的特征，最终与全连接层相连以对图像进行分类，图像类别已事先给定。
- **人脸识别**：CNN 技术不受位置、亮度等特征的影响，它能够从光线不好的图像或者侧脸人像中识别出人脸并进行处理。
- **场景标注**：在场景标注中，每个像素都被标记为所属对象的类别。这里，使用 CNN

在不同层次上对像素进行结合。

- **自然语言处理**：在自然语言处理中，CNN 与词袋模型（bag-of-words）用法类似，但在邮件分类或文本分类等任务中，词袋模型无法胜任。而通过将句子表示为向量矩阵，就可以使用 CNN 完成分类。需要注意是，在这里 CNN 进行一维卷积时，宽度是常量，那么滤波器只需在广度上移动（使用二元文法（bi-gram）时，滤波器广度为 2，使用三元文法（tri-gram）时，广度为 3，依此类推）。

9.1.2　循环神经网络

循环神经网络通过在每个时间步长（time step）上应用递推公式来处理向量 $\boldsymbol{X}$ 的序列。卷积神经网络假设所有的输入都是相互独立的。但是在某些任务中，输入是相互依赖的。例如，在时间序列上预测数据，或根据过去的单词预测句子中的下一个单词等。针对这些需要考虑与过往序列依赖关系的任务，常使用 RNN 进行建模，因为 RNN 提供了更高的准确性。理论上，RNN 可以利用任意长的序列信息，但实际上，它们只能利用最近几步的序列信息。下面的公式解释了 RNN 的原理：

$$h_t = f_W\ (h_{t-1}, \boldsymbol{x}_t)$$

$$h_t = tanh\ (W_{hh}h_{t-1} + W_{xh}\boldsymbol{x}_t)$$

$$= W_{hy}h_t$$

h_t = 新 y_t 状态；

h_W = 以 W 为参数的一些函数

h_{t-1} = 前状态；$\boldsymbol{x}_t$ = 以一定时间步长采样的输入向量

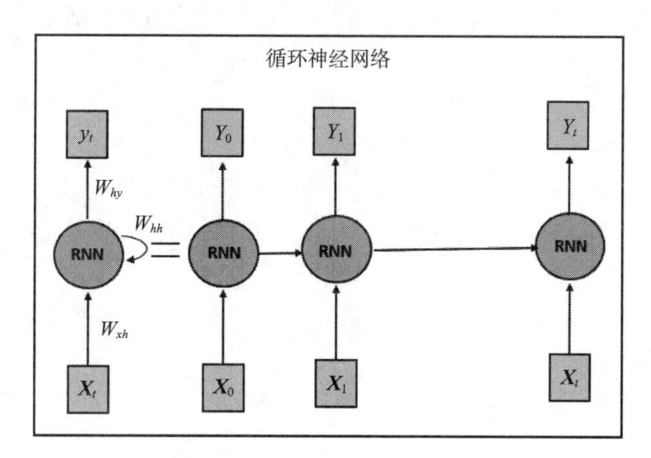

- **RNN 中的梯度消失和梯度爆炸问题**：随着网络层数的增多，梯度消失问题会变得越来越严重。这是因为在每一层都有很多个时间步，每个时间步长都会发生梯度消失问题，且循环权重的求导过程是连续相乘的。因此梯度会迅速消失，梯度爆炸问题同理。这种现象会对 RNN 产生很严重的影响，使得神经网络无法正常训练。梯度爆炸可以通过使用梯度剪裁技术来限制，具体做法是为梯度爆炸设置一个上限，然而梯度消失问题仍然存在。目前，梯度消失问题可以通过使用**长短型记忆网络**（Long Short-term Memory，LSTM）来克服。

- **LSTM**：LSTM 是一个人工神经网络，除常规网络单元外，还包含 LSTM 模块。LSTM 模块包含三个门，输入门、遗忘门和输出门。输入门用来确定何时输入足够重要的信息，遗忘门用来确定应该何时继续记住或忘记上一时刻信息，输出门则用来确定应该何时输出当前时刻信息。

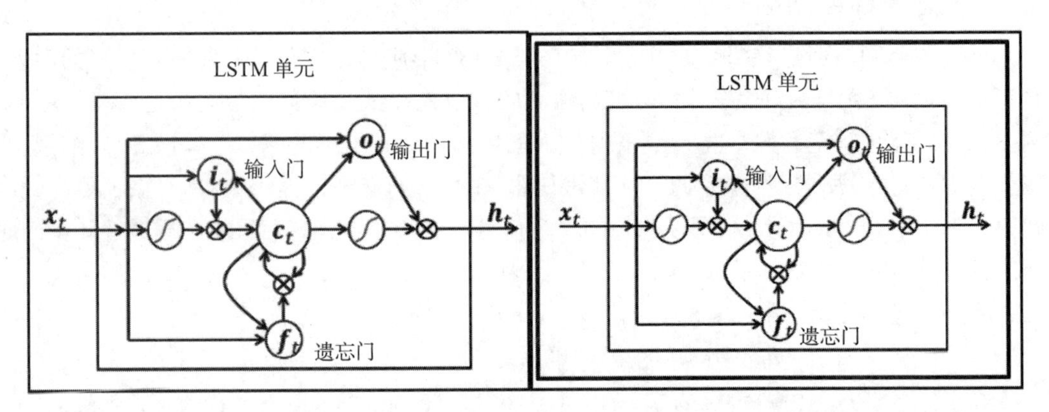

梯度爆炸问题和梯度消失问题在 LSMT 中不会出现，因为它等同于一个加法模型而不是在 RNN 情况下的乘法模型。

RNN 在 NLP 领域中的应用

RNN 已经在很多自然语言处理任务中获得了巨大的成功。特别是，由于克服了梯度消失和梯度爆炸问题，LSTM 成为 RNN 变种中最常用的方式。

- **语言模型**（Language modeling）：在给定一个词序列的情况下预测下一个可能出现的单词。
- **文本生成**（Text generation）：根据一些作者的作品产生新文本。
- **机器翻译**（Machine translation）：将一种语言转换成另一种语言（把英文转换成中文等）。
- **聊天机器人**（Chat bot）：该应用与机器翻译非常类似。不同的是，在聊天机器人中，需要用双方问答对来训练模型。
- **生成图像描述**（Generating an image description）：将 RNN 与 CNN 一起训练来生成图像的标题 / 描述。

9.2　利用深度神经网络对电子邮件进行分类

在本节中，我们将利用深度神经网络对电子邮件进行分类，根据每封电子邮件中出现的单词将其归为 20 个预设类别中的一类。我们从一个简单模型开始，了解深度学习及其在自然语言处理中的应用。

9.2.1　准备工作

我们使用 scikit-learn 中的 20 个新闻组数据集来执行电子邮件的分类任务。该数据集中电子邮件总数量为 18 846（训练集为 11 314，测试集为 7 532），其相应的类别数为 20，如下所示：

```
>>> from sklearn.datasets import fetch_20newsgroups
>>> newsgroups_train = fetch_20newsgroups(subset='train')
>>> newsgroups_test = fetch_20newsgroups(subset='test')
>>> x_train = newsgroups_train.data
>>> x_test = newsgroups_test.data
>>> y_train = newsgroups_train.target
>>> y_test = newsgroups_test.target
>>> print ("List of all 20 categories:")
>>> print (newsgroups_train.target_names)
>>> print ("\n")
>>> print ("Sample Email:")
>>> print (x_train[0])
>>> print ("Sample Target Category:")
>>> print (y_train[0])
>>> print (newsgroups_train.target_names[y_train[0]])
```

下面的屏幕截图显示了第一个训练样本及其目标类别。通过观察第一封电子邮件，我们可以推断，该邮件是在谈论一辆双门跑车，因此我们可以手动将其归类为编号为 8 的汽车类别。

> 由于分类的索引从 0 开始，因此编号为 8 的类别在目标类别中的索引为 7，这与我们的分类结果是一致的。

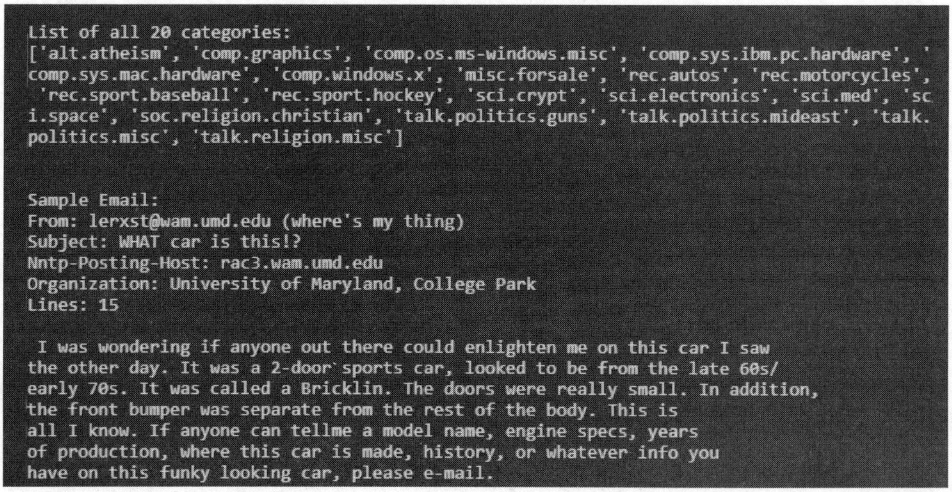

```
List of all 20 categories:
['alt.atheism', 'comp.graphics', 'comp.os.ms-windows.misc', 'comp.sys.ibm.pc.hardware', '
comp.sys.mac.hardware', 'comp.windows.x', 'misc.forsale', 'rec.autos', 'rec.motorcycles',
 'rec.sport.baseball', 'rec.sport.hockey', 'sci.crypt', 'sci.electronics', 'sci.med', 'sc
i.space', 'soc.religion.christian', 'talk.politics.guns', 'talk.politics.mideast', 'talk.
politics.misc', 'talk.religion.misc']


Sample Email:
From: lerxst@wam.umd.edu (where's my thing)
Subject: WHAT car is this!?
Nntp-Posting-Host: rac3.wam.umd.edu
Organization: University of Maryland, College Park
Lines: 15

 I was wondering if anyone out there could enlighten me on this car I saw
the other day. It was a 2-door sports car, looked to be from the late 60s/
early 70s. It was called a Bricklin. The doors were really small. In addition,
the front bumper was separate from the rest of the body. This is
all I know. If anyone can tellme a model name, engine specs, years
of production, where this car is made, history, or whatever info you
have on this funky looking car, please e-mail.
```

```
Thanks,
- IL
   ---- brought to you by your neighborhood Lerxst ----


Sample Target Category:
7
rec.autos
```

9.2.2 如何实现

利用 NLP 技术，我们对数据进行预处理，以获得最终的单词向量来映射垃圾邮件。涉及的主要步骤：

1. 预处理
    1）去除标点符号
    2）分词
    3）将单词转化为小写字母
    4）去除停用词
    5）保留长度至少为 3 的词
    6）提取词干
    7）词性标注
    8）词形还原
2. TF-IDF 向量转换
3. 深度学习模型的训练和测试
4. 模型评估和结果分析

9.2.3 工作原理

所有的预处理步骤都采用了 NLTK 库，因为它包含了所有必要的 NLP 功能：

```
# Used for pre-processing data
>>> import nltk
>>> from nltk.corpus import stopwords
>>> from nltk.stem import WordNetLemmatizer
>>> import string
>>> import pandas as pd
>>> from nltk import pos_tag
>>> from nltk.stem import PorterStemmer
```

为了方便起见，所写的函数（预处理）涵盖了所有步骤。但是，接下来我们将分别解释各部分中的具体步骤：

```
>>> def preprocessing(text):
```

下面的代码行以空字符分割字符串，核查每个字符是否包含标点符号，如果包含标点，那么将用空格代替标点：

```
...     text2 = " ".join("".join([" " if ch in string.punctuation else ch
for ch in text]).split())
```

下面这行代码基于空格把句子切分成词，然后把词放入列表：

```
...     tokens = [word for sent in nltk.sent_tokenize(text2) for word in
nltk.word_tokenize(sent)]
```

下面的代码将所有形式（大写字母、小写字母以及大小写字母相结合）的词都转化为小写字母，用于减少语料中的重复现象：

```
...     tokens = [word.lower() for word in tokens]
```

如前文所述，停用词是在理解句子时携带语义较少的词，它们通常用来连接前后句子等。运用下面的代码可以去除停用词：

```
...     stopwds = stopwords.words('english')
...     tokens = [token for token in tokens if token not in stopwds]
```

在下面的代码中，只保留长度大于 3 的单词，因为长度短的词基本不包含任何语义：

```
...     tokens = [word for word in tokens if len(word)>=3]
```

对于有多余后缀的词，使用 Porter 词干提取器提取词干：

```
...     stemmer = PorterStemmer()
...     tokens = [stemmer.stem(word) for word in tokens]
```

词性标注是词形还原的前提条件，因为只有在了解了一个词是动词或者是名词等词性的前提下，才能将其缩减为词根：

```
...     tagged_corpus = pos_tag(tokens)
```

pos_tag 函数返回 4 种名词形式和 6 种动词形式。NN（名词，普通名词，单数形式），NNP（名词，专有名词，单数形式），NNPS（名词，专有名词，复数形式），NNS（名词，普通名词，复数形式），VB（动词，基本形式），VBD（动词，过去时态），VBG（动词，现在分词），VBN（动词，过去分词），VBP（动词，现在时态，非第三人称单数），VBZ（动词，现在时态，第三人称单数）：

```
...     Noun_tags = ['NN','NNP','NNPS','NNS']
...    Verb_tags = ['VB','VBD','VBG','VBN','VBP','VBZ']
...     lemmatizer = WordNetLemmatizer()
```

由于 pos_tag 函数返回的词性标签和 lemmatize 函数的输入参数不匹配，所以定义以下 prat_lemmatize 函数来进行转换，如果任意一个词的词性被归为名词或动词，那么 lemmatize 函数相应输入 n 或者 v 参数：

```
...     def prat_lemmatize(token,tag):
...       if tag in Noun_tags:
...         return lemmatizer.lemmatize(token,'n')
...       elif tag in Verb_tags:
```

```
...           return lemmatizer.lemmatize(token,'v')
...         else:
...           return lemmatizer.lemmatize(token,'n')
```

在执行了分词和其他所有操作之后，利用下面的函数将数据连接成字符串的形式：

```
...     pre_proc_text =  " ".join([prat_lemmatize(token,tag) for token,tag
in tagged_corpus])
...     return pre_proc_text
```

对训练集和测试集进行预处理：

```
>>> x_train_preprocessed = []
>>> for i in x_train:
... x_train_preprocessed.append(preprocessing(i))
>>> x_test_preprocessed = []
>>> for i in x_test:
... x_test_preprocessed.append(preprocessing(i))
# building TFIDF vectorizer
>>> from sklearn.feature_extraction.text import TfidfVectorizer
>>> vectorizer = TfidfVectorizer(min_df=2, ngram_range=(1, 2),
stop_words='english', max_features= 10000,strip_accents='unicode',
norm='l2')
>>> x_train_2 = vectorizer.fit_transform(x_train_preprocessed).todense()
>>> x_test_2 = vectorizer.transform(x_test_preprocessed).todense()
```

对数据进行预处理后，将得到的 TF-IDF 向量输入到下面的深度学习代码中：

```
# Deep Learning modules
>>> import numpy as np
>>> from keras.models import Sequential
>>> from keras.layers.core import Dense, Dropout, Activation
>>> from keras.optimizers import Adadelta,Adam,RMSprop
>>> from keras.utils import np_utils
```

上面的 Keras 代码运行后得到的结果如下图所示。Keras 能够以 Theano 作为后端在 Python 环境中运行。6GB 内存的 GPU 安装了额外的库（CuDNN 和 CNMeM）后，运行速度提高了 4 到 5 倍，同时这些库只占用了大约 20% 的内存，因此 6GB 内存中有 80% 的内存可用：

```
Using Theano backend.
WARNING (theano.sandbox.cuda): The cuda backend is deprecated and will be removed in the next release (v0.10).  Please s
witch to the gpuarray backend. You can get more information about how to switch at this URL:
 https://github.com/Theano/Theano/wiki/Converting-to-the-new-gpu-back-end%28gpuarray%29

Using gpu device 0: GeForce GTX 1060 6GB (CNMeM is enabled with initial size: 80.0% of memory, cuDNN 5105)
```

下面的代码解释了深度学习模型的核心部分。可以看出，以下代码分别设置分类类别数（nb_classes）为 20，批尺寸（batch_size）为 64，训练迭代次数（nb_epochs）为 20：

```
# Definition hyper parameters
>>> np.random.seed(1337)
>>> nb_classes = 20
>>> batch_size = 64
>>> nb_epochs = 20
```

下面的代码将 20 个类别转换为 one-hot 编码向量，即创建 20 列，将每个类别相应位置的值设为1，其余位置设为0：

```
>>> Y_train = np_utils.to_categorical(y_train, nb_classes)
```

下面的Keras代码块使用了三个隐藏层（每一层神经元的个数分别为1000、500和50），每层的dropout均为50%，优化算法为Adam：

```
#Deep Layer Model building in Keras
#del model
>>> model = Sequential()
>>> model.add(Dense(1000,input_shape= (10000,)))
>>> model.add(Activation('relu'))
>>> model.add(Dropout(0.5))
>>> model.add(Dense(500))
>>> model.add(Activation('relu'))
>>> model.add(Dropout(0.5))
>>> model.add(Dense(50))
>>> model.add(Activation('relu'))
>>> model.add(Dropout(0.5))
>>> model.add(Dense(nb_classes))
>>> model.add(Activation('softmax'))
>>> model.compile(loss='categorical_crossentropy', optimizer='adam')
>>> print (model.summary())
```

模型体系结构如下所示，输入数据首先从包含10 000个神经元的隐藏层开始，然后依次经过包含1000、500、50和20个神经元的隐藏层，最后对给定的电子邮件进行分类，将其归于20类中的一类：

Layer (type)	Output Shape	Param #
dense_1 (Dense)	(None, 1000)	10001000
activation_1 (Activation)	(None, 1000)	0
dropout_1 (Dropout)	(None, 1000)	0
dense_2 (Dense)	(None, 500)	500500
activation_2 (Activation)	(None, 500)	0
dropout_2 (Dropout)	(None, 500)	0
dense_3 (Dense)	(None, 50)	25050
activation_3 (Activation)	(None, 50)	0
dropout_3 (Dropout)	(None, 50)	0
dense_4 (Dense)	(None, 20)	1020
activation_4 (Activation)	(None, 20)	0

Total params: 10,527,570.0
Trainable params: 10,527,570.0
Non-trainable params: 0.0

按照给定的参数进行模型训练：

```
# Model Training
>>> model.fit(x_train_2, Y_train, batch_size=batch_size,
epochs=nb_epochs,verbose=1)
```

该模型已经训练了 20 轮，每一轮耗费约 2 秒。损失从 1.9281 减小到了 0.0241。若使用 CPU 训练，每一轮花费的时间可能会增加，因为 GPU 可以通过数以千计的线程实现大规模并行计算：

```
Epoch 1/20
11314/11314 [==============================] - 2s - loss: 1.9281
Epoch 2/20
11314/11314 [==============================] - 2s - loss: 0.5844
Epoch 3/20
11314/11314 [==============================] - 2s - loss: 0.2854
Epoch 4/20
11314/11314 [==============================] - 2s - loss: 0.1709
Epoch 17/20
11314/11314 [==============================] - 2s - loss: 0.0218
Epoch 18/20
11314/11314 [==============================] - 2s - loss: 0.0217
Epoch 19/20
11314/11314 [==============================] - 2s - loss: 0.0229
Epoch 20/20
11314/11314 [==============================] - 2s - loss: 0.0241
Out[13]: <keras.callbacks.History at 0x1701c0f60>
```

最后，在训练集和测试集上进行预测，并计算准确率和召回率：

```
#Model Prediction
>>> y_train_predclass =
model.predict_classes(x_train_2,batch_size=batch_size)
>>> y_test_predclass =
model.predict_classes(x_test_2,batch_size=batch_size)
>>> from sklearn.metrics import accuracy_score,classification_report
>>> print ("\n\nDeep Neural Network - Train
accuracy:"),(round(accuracy_score( y_train, y_train_predclass),3))
>>> print ("\nDeep Neural Network - Test accuracy:"),(round(accuracy_score(
y_test,y_test_predclass),3))
>>> print ("\nDeep Neural Network - Train Classification Report")
>>> print (classification_report(y_train,y_train_predclass))
>>> print ("\nDeep Neural Network - Test Classification Report")
>>> print (classification_report(y_test,y_test_predclass))
```

如下图所示，该分类器在训练集上准确率高达 99.9%，在测试集上准确率为 80.7%：

```
Deep Neural Network  - Train accuracy: 0.999

Deep Neural Network  - Test accuracy: 0.807

Deep Neural Network  - Train Classification Report
             precision    recall  f1-score   support

          0       1.00      1.00      1.00       480
          1       0.99      1.00      1.00       584
          2       1.00      1.00      1.00       591
          3       1.00      1.00      1.00       590
          4       1.00      1.00      1.00       578
          5       1.00      1.00      1.00       593
          6       1.00      1.00      1.00       585
          7       1.00      1.00      1.00       594
          8       1.00      1.00      1.00       598
          9       1.00      1.00      1.00       597
         10       1.00      1.00      1.00       600
         11       1.00      1.00      1.00       595
         12       1.00      1.00      1.00       591
```

```
         13       1.00      1.00      1.00       594
         14       1.00      1.00      1.00       593
         15       1.00      1.00      1.00       599
         16       1.00      1.00      1.00       546
         17       1.00      1.00      1.00       564
         18       1.00      1.00      1.00       465
         19       1.00      1.00      1.00       377

avg / total       1.00      1.00      1.00     11314


Deep Neural Network  - Test Classification Report
             precision    recall  f1-score   support

          0       0.74      0.73      0.74       319
          1       0.61      0.75      0.67       389
          2       0.74      0.69      0.71       394
          3       0.71      0.67      0.69       392
          4       0.76      0.78      0.77       385
          5       0.86      0.76      0.81       395
          6       0.85      0.80      0.82       390
          7       0.89      0.84      0.86       396
          8       0.94      0.91      0.92       398
          9       0.91      0.89      0.90       397
         10       0.94      0.97      0.96       399
         11       0.92      0.91      0.91       396
         12       0.64      0.75      0.69       393
         13       0.92      0.80      0.86       396
         14       0.92      0.89      0.91       394
         15       0.81      0.89      0.85       398
         16       0.75      0.89      0.81       364
         17       0.94      0.82      0.88       376
         18       0.77      0.64      0.70       310
         19       0.55      0.65      0.60       251

avg / total       0.81      0.81      0.81      7532
```

9.3　使用一维卷积网络进行 IMDB 情感分类

在本节中，我们将使用 Keras IMDB 电影评论情感数据，这些数据标有情感标签（正向 / 负向）。首先，我们要对评论做预处理。接着将每一条评论编码成词索引（整数）序列。最后，我们将用下面的代码对其进行解码。

9.3.1　准备工作

来自 Keras 的 IMDB 数据集包含了一个词集和与其相对应的情感标签。下面将对数据进行预处理：

```
>>> import pandas as pd
>>> from keras.preprocessing import sequence
>>> from keras.models import Sequential
>>> from keras.layers import Dense, Dropout, Activation
>>> from keras.layers import Embedding
>>> from keras.layers import Conv1D, GlobalMaxPooling1D
>>> from keras.datasets import imdb
>>> from sklearn.metrics import accuracy_score,classification_report
```

在下面的这组参数中，我们设置最大特征数为 6000，单个句子的最大长度为 400：

```
# set parameters:
>>> max_features = 6000
>>> max_length = 400
>>> (x_train, y_train), (x_test, y_test) =
imdb.load_data(num_words=max_features)
>>> print(len(x_train), 'train observations')
>>> print(len(x_test), 'test observations')
```

IMDB 数据集拥有相同数量的训练样例和测试样例，我们将在 25 000 条训练数据上建立模型，然后用 25 000 条测试数据对模型进行测试。以下截图显示了训练集和测试集的数量：

```
25000 train observations
25000 test observations
```

下面的代码为每个词创建了一个字典映射，使每个词唯一地对应一个整数索引值：

```
# Creating numbers to word mapping
>>> wind = imdb.get_word_index()
>>> revind = dict((v,k) for k,v in wind.iteritems())
>>> print (x_train[0])
>>> print (y_train[0])
```

因为计算机不能直接处理字母、单词等数据，而只能对数字进行理解和处理，所以我们需要把英文句子转换成数字序列来进行处理。如下面的截图所示，我们将第一条数据通过字典映射转换为对应的数字序列，其中不再包含任何英文单词：

```
[1, 14, 22, 16, 43, 530, 973, 1622, 1385, 65, 458, 4468, 66, 3941, 4, 173, 36, 256, 5, 25, 100, 43, 838, 112, 50, 670, 2
, 9, 35, 480, 284, 5, 150, 4, 172, 112, 167, 2, 336, 385, 39, 4, 172, 4536, 1111, 17, 546, 38, 13, 447, 4, 192, 50, 16,
6, 147, 2025, 19, 14, 22, 4, 1920, 4613, 469, 4, 22, 71, 87, 12, 16, 43, 530, 38, 76, 15, 13, 1247, 4, 22, 17, 515, 17,
12, 16, 626, 18, 2, 5, 62, 386, 12, 8, 316, 8, 106, 5, 4, 2223, 5244, 16, 480, 66, 3785, 33, 4, 130, 12, 16, 38, 619, 5,
 25, 124, 51, 36, 135, 48, 25, 1415, 33, 6, 22, 12, 215, 28, 77, 52, 5, 14, 407, 16, 82, 2, 8, 4, 107, 117, 5952, 15, 25
6, 4, 2, 7, 3766, 5, 723, 36, 71, 43, 530, 476, 26, 400, 317, 46, 7, 4, 2, 1029, 13, 104, 88, 4, 381, 15, 297, 98, 32, 2
071, 56, 26, 141, 6, 194, 2, 18, 4, 226, 22, 21, 134, 476, 26, 480, 5, 144, 30, 5535, 18, 51, 36, 28, 224, 92, 25, 104,
4, 226, 65, 16, 38, 1334, 88, 12, 16, 283, 5, 16, 4472, 113, 103, 32, 15, 16, 5345, 19, 178, 32]
1
```

下面的代码显示了如何使用创建的逆映射字典进行解码：

```
>>> def decode(sent_list):
... new_words = []
... for i in sent_list:
... new_words.append(revind[i])
... comb_words = " ".join(new_words)
... return comb_words
>>> print (decode(x_train[0]))
```

将数字序列转换为文本序列的结果如下图所示。这里，我们使用字典将整数形式转化成文本形式：

```
the as you with out themselves powerful lets loves their becomes reaching had journalist of lot from anyone to have after
 out atmosphere never more room and it so heart shows to years of every never going and help moments or of every chest vi
sual movie except her was several of enough more with is now current film as you of mine potentially unfortunately of you
 than him that with out themselves her get for was camp of you movie sometimes movie that with scary but and to story won
derful that in seeing in character to of 70s musicians with heart had shadows they of here that with her serious to have
does when from why what have critics they is you that isn't one will very to as itself with other and in of seen over lan
ded for anyone of and br show's to whether from than out themselves history he name half some br of and odd was two most
of mean for 1 any an boat she he should is thought and but of script you not while history he heart to real at barrel but
 when from one bit then have two of script their with her nobody most that with wasn't to with armed acting watch an for
with heartfelt film want an
```

9.3.2　如何实现

所涉及的主要步骤如下：

1. 预处理。在这个阶段，我们通过序列填充将所有的数据整合为一个固定的维度，这样的数据才具备可计算性并能提高运行速度。

2. 一维 CNN 模型的构建和验证。

3. 模型评估。

9.3.3　工作原理

下面的代码执行了填充操作，通过添加额外的句子使原始句子的长度达到 400 个单词的最大限值。这样做的好处是使数据长度保持一致，并且更适用于神经网络计算：

```
#Pad sequences for computational efficiency
>>> x_train = sequence.pad_sequences(x_train, maxlen=max_length)
>>> x_test = sequence.pad_sequences(x_test, maxlen=max_length)
>>> print('x_train shape:', x_train.shape)
>>> print('x_test shape:', x_test.shape)
```

```
x_train shape: (25000L, 400L)
x_test shape: (25000L, 400L)
```

下面的深度学习代码用于创建一维 CNN 模型，其中运用了 Keras 框架：

```
# Deep Learning architecture parameters
>>> batch_size = 32
>>> embedding_dims = 60
>>> num_kernels = 260
>>> kernel_size = 3
>>> hidden_dims = 300
>>> epochs = 3
# Building the model
>>> model = Sequential()
>>> model.add(Embedding(max_features,embedding_dims, input_length=
max_length))
>>> model.add(Dropout(0.2))
>>> model.add(Conv1D(num_kernels,kernel_size, padding='valid',
activation='relu', strides=1))
>>> model.add(GlobalMaxPooling1D())
>>> model.add(Dense(hidden_dims))
>>> model.add(Dropout(0.5))
>>> model.add(Activation('relu'))
>>> model.add(Dense(1))
>>> model.add(Activation('sigmoid'))
>>> model.compile(loss='binary_crossentropy',optimizer='adam',
metrics=['accuracy'])
>>> print (model.summary())
```

下面的截图完整地展示了整个模型概况，给出了模型各层的维数及其相应的神经元个数，这些将直接影响从输入数据到最终目标变量（无论是 0 还是 1）过程中参与计算的参数数量。因此，在网络的最后一层使用了一个稠密层（dense layer）：

```
Layer (type)                 Output Shape              Param #
=================================================================
embedding_1 (Embedding)      (None, 400, 60)           360000
_________________________________________________________________
dropout_1 (Dropout)          (None, 400, 60)           0
_________________________________________________________________
conv1d_1 (Conv1D)            (None, 398, 260)          47060
_________________________________________________________________
global_max_pooling1d_1 (Glob (None, 260)               0
_________________________________________________________________
dense_1 (Dense)              (None, 300)               78300
_________________________________________________________________
dropout_2 (Dropout)          (None, 300)               0
_________________________________________________________________
activation_1 (Activation)    (None, 300)               0
_________________________________________________________________
dense_2 (Dense)              (None, 1)                 301
_________________________________________________________________
activation_2 (Activation)    (None, 1)                 0
=================================================================
Total params: 485,661.0
Trainable params: 485,661.0
Non-trainable params: 0.0
_________________________________________________________________
None
```

下面的代码对训练数据执行模型拟合操作，训练数据中的 X 和 Y 变量都被用于分批训练数据：

```
>>> model.fit(x_train, y_train,batch_size=batch_size,epochs=epochs,
validation_split=0.2)
```

该模型已经训练了 3 轮，每一轮在 GPU 上消耗 5 秒。当观察更多次迭代结果时，我们发现尽管在训练集上的准确率在持续上升，但是在验证集上的准确率却在下降。我们称这种现象为模型过拟合。在一定范围内，迭代次数的增加可以提高模型准确率。由于不断增加迭代次数的同时会出现过拟合情况，因此我们在适度增加迭代次数的同时也要寻找其他方法来提高模型的准确率。例如，我们可以考虑扩展体系结构规模等方法。鼓励读者尝试各种组合：

```
Train on 20000 samples, validate on 5000 samples
Epoch 1/3
20000/20000 [==============================] - 5s - loss: 0.4321 - acc: 0.7672 - val_loss: 0.2896 - val_acc: 0.8750
Epoch 2/3
20000/20000 [==============================] - 5s - loss: 0.2498 - acc: 0.9001 - val_loss: 0.2890 - val_acc: 0.8802
Epoch 3/3
20000/20000 [==============================] - 5s - loss: 0.1635 - acc: 0.9397 - val_loss: 0.2875 - val_acc: 0.8836
      <keras.callbacks.History at 0x1668d4f28>
```

下面的代码用于预测训练集和测试集的类别：

```
#Model Prediction
>>> y_train_predclass =
model.predict_classes(x_train,batch_size=batch_size)
>>> y_test_predclass = model.predict_classes(x_test,batch_size=batch_size)
>>> y_train_predclass.shape = y_train.shape
>>> y_test_predclass.shape = y_test.shape

# Model accuracies and metrics calculation
```

```
>>> print (("\n\nCNN 1D - Train accuracy:"),(round(accuracy_score(y_train,
y_train_predclass),3)))
>>> print ("\nCNN 1D of Training data\n",classification_report(y_train,
y_train_predclass))
>>> print ("\nCNN 1D - Train Confusion Matrix\n\n",pd.crosstab(y_train,
y_train_predclass,rownames = ["Actuall"],colnames = ["Predicted"]))
>>> print (("\nCNN 1D - Test accuracy:"),(round(accuracy_score(y_test,
y_test_predclass),3)))
>>> print ("\nCNN 1D of Test data\n",classification_report(y_test,
y_test_predclass))
>>> print ("\nCNN 1D - Test Confusion Matrix\n\n",pd.crosstab(y_test,
y_test_predclass,rownames = ["Actuall"],colnames = ["Predicted"]))
```

下面的截图展示了测试模型性能的多种可评价指标。从结果来看，在训练集上的准确率高达 96%；然而，在测试集上的准确度较低，只有 88.2%。这种情况可能是由于模型过度拟合造成的：

```
CNN 1D  - Train accuracy: 0.96

CNN 1D of Training data
             precision    recall  f1-score   support

          0       0.97      0.95      0.96     12500
          1       0.95      0.97      0.96     12500

avg / total       0.96      0.96      0.96     25000


CNN 1D - Train Confusion Matrix

 Predicted      0      1
Actuall
0           11825    675
1             319  12181

CNN 1D  - Test accuracy: 0.882

CNN 1D of Test data
             precision    recall  f1-score   support

          0       0.90      0.86      0.88     12500
          1       0.86      0.91      0.89     12500

avg / total       0.88      0.88      0.88     25000


CNN 1D - Test Confusion Matrix

 Predicted      0      1
Actuall
0           10689   1811
1            1139  11361
```

9.4 基于双向 LSTM 的 IMDB 情感分类模型

在本节中，我们同样使用 IMDB 情绪数据来说明 CNN 和 RNN 方法在准确率等方面的差异。数据预处理步骤保持不变，只是模型的架构有所不同。

9.4.1 准备工作

来自 Keras 的 IMDB 数据集中包含了一个词集合及其相应的情感标签。下面将对数据进行预处理：

```
>>> from __future__ import print_function
>>> import numpy as np
>>> import pandas as pd
>>> from keras.preprocessing import sequence
>>> from keras.models import Sequential
>>> from keras.layers import Dense, Dropout, Embedding, LSTM, Bidirectional
>>> from keras.datasets import imdb
>>> from sklearn.metrics import accuracy_score,classification_report

# Max features are limited
>>> max_features = 15000
>>> max_len = 300
>>> batch_size = 64

# Loading data
>>> (x_train, y_train), (x_test, y_test) =
imdb.load_data(num_words=max_features)
>>> print(len(x_train), 'train observations')
>>> print(len(x_test), 'test observations')
```

9.4.2 如何实现

所涉及的主要步骤如下：

1. 预处理。在这个阶段，我们通过序列填充将所有的数据整合为一个固定的维度，这样的数据才具备可计算性并能提高运行速度。

2. LSTM 模型的构建和验证。

3. 模型评估。

9.4.3 工作原理

```
# Pad sequences for computational efficiently
>>> x_train_2 = sequence.pad_sequences(x_train, maxlen=max_len)
>>> x_test_2 = sequence.pad_sequences(x_test, maxlen=max_len)
>>> print('x_train shape:', x_train_2.shape)
>>> print('x_test shape:', x_test_2.shape)
>>> y_train = np.array(y_train)
>>> y_test = np.array(y_test)
```

接下来的深度学习代码利用 Keras 库创建了一个双向 LSTM 模型。

双向 LSTM 网络具有前向和后向的双向连接，这使处于句子中的单词可以同时与其左右词汇产生连接：

```
# Model Building
>>> model = Sequential()
>>> model.add(Embedding(max_features, 128, input_length=max_len))
>>> model.add(Bidirectional(LSTM(64)))
>>> model.add(Dropout(0.5))
>>> model.add(Dense(1, activation='sigmoid'))
>>> model.compile('adam', 'binary_crossentropy', metrics=['accuracy'])
# Print model architecture
>>> print (model.summary())
```

下面的截图显示了双向 LSTM 网络的架构。首先使用嵌入层（embedding layer）将维数降低到 128，然后是双向 LSTM 层，最后使用稠密层（dense layer）将情感分类为 0 或 1：

```
_________________________________________________________________
Layer (type)                 Output Shape              Param #
=================================================================
embedding_1 (Embedding)      (None, 300, 128)          1920000
_________________________________________________________________
bidirectional_1 (Bidirection (None, 128)               98816
_________________________________________________________________
dropout_1 (Dropout)          (None, 128)               0
_________________________________________________________________
dense_1 (Dense)              (None, 1)                 129
=================================================================
Total params: 2,018,945.0
Trainable params: 2,018,945.0
Non-trainable params: 0.0
_________________________________________________________________
None
```

下面的代码用于训练数据：

```
#Train the model
>>> model.fit(x_train_2, y_train,batch_size=batch_size,epochs=4,
validation_split=0.2)
```

因为 LSTM 不容易在 GPU（4x 到 5x）上实现并行运算，而 CNN（100x）可以进行大规模并行运算，因此，与 CNN 相比，LSTM 模型需要耗费更长的时间。通过对每一轮迭代结果的观察，我们发现随着模型迭代次数的增加，模型在训练集上的准确率有所提高但是验证准确率却在下降。这种情况表明该模型处于过度拟合状态：

```
Train on 20000 samples, validate on 5000 samples
Epoch 1/4
20000/20000 [==============================] - 205s - loss: 0.4366 - acc: 0.7936 - val_loss: 0.3239 - val_acc: 0.8656
Epoch 2/4
20000/20000 [==============================] - 205s - loss: 0.2352 - acc: 0.9128 - val_loss: 0.3779 - val_acc: 0.8676
Epoch 3/4
20000/20000 [==============================] - 205s - loss: 0.1661 - acc: 0.9426 - val_loss: 0.3661 - val_acc: 0.8664
Epoch 4/4
20000/20000 [==============================] - 203s - loss: 0.1102 - acc: 0.9626 - val_loss: 0.3887 - val_acc: 0.8638
<keras.callbacks.History at 0x167954d30>
```

以下代码用于在训练集和测试集上预测情感类别：

```
#Model Prediction
>>> y_train_predclass = model.predict_classes(x_train_2,batch_size=1000)
>>> y_test_predclass = model.predict_classes(x_test_2,batch_size=1000)
>>> y_train_predclass.shape = y_train.shape
>>> y_test_predclass.shape = y_test.shape
```

```
# Model accuracies and metrics calculation
>>> print (("\n\nLSTM Bidirectional Sentiment Classification - Train
accuracy:"),(round(accuracy_score(y_train,y_train_predclass),3)))
>>> print ("\nLSTM Bidirectional Sentiment Classification of Training
data\n",classification_report(y_train, y_train_predclass))
>>> print ("\nLSTM Bidirectional Sentiment Classification - Train Confusion
Matrix\n\n",pd.crosstab(y_train, y_train_predclass,rownames =
["Actuall"],colnames = ["Predicted"]))
>>> print (("\nLSTM Bidirectional Sentiment Classification - Test
accuracy:"),(round(accuracy_score(y_test,y_test_predclass),3)))
>>> print ("\nLSTM Bidirectional Sentiment Classification of Test
data\n",classification_report(y_test, y_test_predclass))
>>> print ("\nLSTM Bidirectional Sentiment Classification - Test Confusion
Matrix\n\n",pd.crosstab(y_test, y_test_predclass,rownames =
["Actuall"],colnames = ["Predicted"]))
```

观察下面的截图我们发现，与 CNN 模型相比，LSTM 模型的测试准确率较低。但是，如果我们仔细调整模型参数，就可以在 RNN 模型上获得比 CNN 模型更高的准确率：

```
LSTM Bidirectional Sentiment Classification  - Train accuracy: 0.957

LSTM Bidirectional Sentiment Classification of Training data
             precision    recall  f1-score   support

          0       0.95      0.97      0.96     12500
          1       0.97      0.94      0.96     12500

avg / total       0.96      0.96      0.96     25000


LSTM Bidirectional Sentiment Classification - Train Confusion Matrix

 Predicted      0      1
Actuall
0          12124    376
1            700  11800

LSTM Bidirectional Sentiment Classification  - Test accuracy: 0.856

LSTM Bidirectional Sentiment Classification of Test data
             precision    recall  f1-score   support

          0       0.83      0.89      0.86     12500
          1       0.88      0.82      0.85     12500

avg / total       0.86      0.86      0.86     25000


LSTM Bidirectional Sentiment Classification - Test Confusion Matrix

 Predicted      0      1
Actuall
0          11140   1360
1           2242  10258
```

9.5 利用词向量实现高维词在二维空间的可视化

在本节中，我们将利用深度神经网络实现单词从高维空间到二维空间的可视化。

9.5.1 准备工作

爱丽丝梦游仙境（Alice in Wonderland）数据集可用于单词抽取，结合稠密网络可实现其单词的可视化。这与编码器—解码器架构类似：

```
>>> from __future__ import print_function
>>> import os
""" First change the following directory link to where all input files do
exist """
>>> os.chdir("C:\\Users\\prata\\Documents\\book_codes\\NLP_DL")
>>> import nltk
>>> from nltk.corpus import stopwords
>>> from nltk.stem import WordNetLemmatizer
>>> from nltk import pos_tag
>>> from nltk.stem import PorterStemmer
>>> import string
>>> import numpy as np
>>> import pandas as pd
>>> import random
>>> from sklearn.model_selection import train_test_split
>>> from sklearn.preprocessing import OneHotEncoder
>>> import matplotlib.pyplot as plt
>>> def preprocessing(text):
... text2 = " ".join("".join([" " if ch in string.punctuation else ch for
ch in text]).split())
... tokens = [word for sent in nltk.sent_tokenize(text2) for word in
nltk.word_tokenize(sent)]
... tokens = [word.lower() for word in tokens]
... stopwds = stopwords.words('english')
... tokens = [token for token in tokens if token not in stopwds]
... tokens = [word for word in tokens if len(word)>=3]
... stemmer = PorterStemmer()
... tokens = [stemmer.stem(word) for word in tokens]
... tagged_corpus = pos_tag(tokens)
... Noun_tags = ['NN','NNP','NNPS','NNS']
... Verb_tags = ['VB','VBD','VBG','VBN','VBP','VBZ']
... lemmatizer = WordNetLemmatizer()
... def prat_lemmatize(token,tag):
... if tag in Noun_tags:
... return lemmatizer.lemmatize(token,'n')
... elif tag in Verb_tags:
... return lemmatizer.lemmatize(token,'v')
... else:
... return lemmatizer.lemmatize(token,'n')
... pre_proc_text = " ".join([prat_lemmatize(token,tag) for token,tag in
tagged_corpus])
... return pre_proc_text
>>> lines = []
```

```
>>> fin = open("alice_in_wonderland.txt", "rb")
>>> for line in fin:
... line = line.strip().decode("ascii", "ignore").encode("utf-8")
... if len(line) == 0:
... continue
... lines.append(preprocessing(line))
>>> fin.close()
```

9.5.2 如何实现

所涉及的主要步骤如下：

1. 预处理。构建 skip-gram 模型，该模型通过给定中心（当前）词来预测左边和右边的词（上下文）。

2. 在特征工程中应用 one—hot 编码。

3. 运用编码器—解码器架构构建模型。

4. 使用编码器架构创建二维特征，实现测试集的可视化。

9.5.3 工作原理

下面的代码用于创建字典，字典是从词到索引的映射，也可以是从索引到词的映射。我们知道，模型不能处理字符或单词形式的输入。因此，我们需要把单词转化为数字索引序列（最好采用整数映射），一旦运用神经网络模型进行计算，那么从数字索引到词的映射可以使其可视化。collections 库的 Counter 函数可用于有效创建字典：

```
>>> import collections
>>> counter = collections.Counter()
>>> for line in lines:
... for word in nltk.word_tokenize(line):
... counter[word.lower()]+=1
>>> word2idx = {w:(i+1) for i,(w,_) in enumerate(counter.most_common())}
>>> idx2word = {v:k for k,v in word2idx.items()}
```

下面的代码用于实现词到整数的映射并从 embedding 中抽取三元文法（tri-gram）特征。若要使 skip-gran 方法在测试阶段能通过中心词正确预测左右相邻词的话，则需将中心词与左右相邻词相连接在一起进行训练：

```
>>> xs = []
>>> ys = []
>>> for line in lines:
... embedding = [word2idx[w.lower()] for w in nltk.word_tokenize(line)]
... triples = list(nltk.trigrams(embedding))
... w_lefts = [x[0] for x in triples]
... w_centers = [x[1] for x in triples]
... w_rights = [x[2] for x in triples]
... xs.extend(w_centers)
... ys.extend(w_lefts)
... xs.extend(w_centers)
... ys.extend(w_rights)
```

下面的代码中字典长度与词表大小相等。此外，我们还可以基于用户要求自定义词表大小：

```
>>> print (len(word2idx))
>>> vocab_size = len(word2idx)+1
```

根据词表大小，下面的代码将所有的自变量和因变量都转化为向量表示，其中行数为词数，列数为词表大小。神经网络模型将输入和输出变量映射到向量空间：

```
>>> ohe = OneHotEncoder(n_values=vocab_size)
>>> X = ohe.fit_transform(np.array(xs).reshape(-1, 1)).todense()
>>> Y = ohe.fit_transform(np.array(ys).reshape(-1, 1)).todense()
>>> Xtrain, Xtest, Ytrain, Ytest,xstr,xsts = train_test_split(X, Y,xs,
test_size=0.3, random_state=42)
>>> print(Xtrain.shape, Xtest.shape, Ytrain.shape, Ytest.shape)
```

在总共的 13 868 个观测样本中，训练集和测试集分别占 70% 和 30%，即训练集和测试集数量分别为 9707 和 4161，如下面的截图所示：

```
(9707L, 1787L) (4161L, 1787L) (9707L, 1787L) (4161L, 1787L)
```

下面是使用 Keras 框架编写的深度学习模型的核心部分。此代码是收敛–发散（convergent-divergent）的，初始化时对输入词维数的压缩是为了与要求输出的格式相匹配。这样做的话，网络第二层维度将降至二维。模型训练完成后，我们将提取第二层，用于对测试数据进行预测。实际上这与传统的编码器–解码器架构类似：

```
>>> from keras.layers import Input,Dense,Dropout
>>> from keras.models import Model
>>> np.random.seed(42)
>>> BATCH_SIZE = 128
>>> NUM_EPOCHS = 20
>>> input_layer = Input(shape = (Xtrain.shape[1],),name="input")
>>> first_layer = Dense(300,activation='relu',name = "first")(input_layer)
>>> first_dropout = Dropout(0.5,name="firstdout")(first_layer)
>>> second_layer = Dense(2,activation='relu',name="second") (first_dropout)
>>> third_layer = Dense(300,activation='relu',name="third") (second_layer)
>>> third_dropout = Dropout(0.5,name="thirdout")(third_layer)
>>> fourth_layer = Dense(Ytrain.shape[1],activation='softmax',name =
"fourth")(third_dropout)
>>> history = Model(input_layer,fourth_layer)
>>> history.compile(optimizer = "rmsprop",loss= "categorical_crossentropy",
metrics=["accuracy"])
```

下面的代码用于训练模型：

```
>>> history.fit(Xtrain, Ytrain, batch_size=BATCH_SIZE,epochs=NUM_EPOCHS,
verbose=1,validation_split = 0.2)
```

仔细观察下面的截图，我们发现训练集和验证集上的最佳准确率不超过 6%，这是由受限的数据规模和深度学习模型架构造成的。为了使该模型真正发挥作用，我们至少需要千兆字节的数据和大型的架构，模型也需要长时间的训练。由于实际的限制和可视化目的，我们仅进行了 20 次迭代训练。但是，我们鼓励读者尝试各种组合，以提高模型准确率：

```
Train on 7765 samples, validate on 1942 samples
Epoch 1/20
7765/7765 [==============================] - 0s - loss: 6.9000 - acc: 0.0382 - val_loss: 6.4369 - val_acc: 0.0479
Epoch 2/20
7765/7765 [==============================] - 0s - loss: 6.3954 - acc: 0.0426 - val_loss: 6.4417 - val_acc: 0.0479
Epoch 3/20
7765/7765 [==============================] - 0s - loss: 6.3458 - acc: 0.0434 - val_loss: 6.4712 - val_acc: 0.0479
Epoch 4/20
7765/7765 [==============================] - 0s - loss: 6.3264 - acc: 0.0438 - val_loss: 6.4962 - val_acc: 0.0479
Epoch 5/20
7765/7765 [==============================] - 0s - loss: 6.3099 - acc: 0.0439 - val_loss: 6.5160 - val_acc: 0.0479
Epoch 16/20
7765/7765 [==============================] - 0s - loss: 6.1590 - acc: 0.0487 - val_loss: 6.6237 - val_acc: 0.0541
Epoch 17/20
7765/7765 [==============================] - 0s - loss: 6.1417 - acc: 0.0498 - val_loss: 6.6250 - val_acc: 0.0530
Epoch 18/20
7765/7765 [==============================] - 0s - loss: 6.1173 - acc: 0.0502 - val_loss: 6.6496 - val_acc: 0.0530
Epoch 19/20
7765/7765 [==============================] - 0s - loss: 6.1029 - acc: 0.0501 - val_loss: 6.6488 - val_acc: 0.0541
Epoch 20/20
7765/7765 [==============================] - 0s - loss: 6.0869 - acc: 0.0509 - val_loss: 6.6308 - val_acc: 0.0515
    <keras.callbacks.History at 0x197984550>
```

```
# Extracting Encoder section of the Model for prediction of latent
variables
>>> encoder = Model(history.input,history.get_layer("second").output)

# Predicting latent variables with extracted Encoder model
>>> reduced_X = encoder.predict(Xtest)
Converting the outputs into Pandas data frame structure for better
representation
>>> final_pdframe = pd.DataFrame(reduced_X)
>>> final_pdframe.columns = ["xaxis","yaxis"]
>>> final_pdframe["word_indx"] = xsts
>>> final_pdframe["word"] = final_pdframe["word_indx"].map(idx2word)
>>> rows = random.sample(final_pdframe.index, 100)
>>> vis_df = final_pdframe.ix[rows]
>>> labels = list(vis_df["word"]);xvals = list(vis_df["xaxis"])
>>> yvals = list(vis_df["yaxis"])

#in inches
>>> plt.figure(figsize=(8, 8))
>>> for i, label in enumerate(labels):
... x = xvals[i]
... y = yvals[i]
... plt.scatter(x, y)
... plt.annotate(label,xy=(x, y),xytext=(5, 2),textcoords='offset points',
ha='right',va='bottom')
>>> plt.xlabel("Dimension 1")
>>> plt.ylabel("Dimension 2")
>>> plt.show()
```

以下图像可视化了二维空间中单词的分布。一些单词相对于其他词来说相互之间更加靠近，这表示这些词拥有更紧密的关系。例如，单词 never、ever 和 ask 相互之间非常接近，它们在语义上也有更紧密的联系。

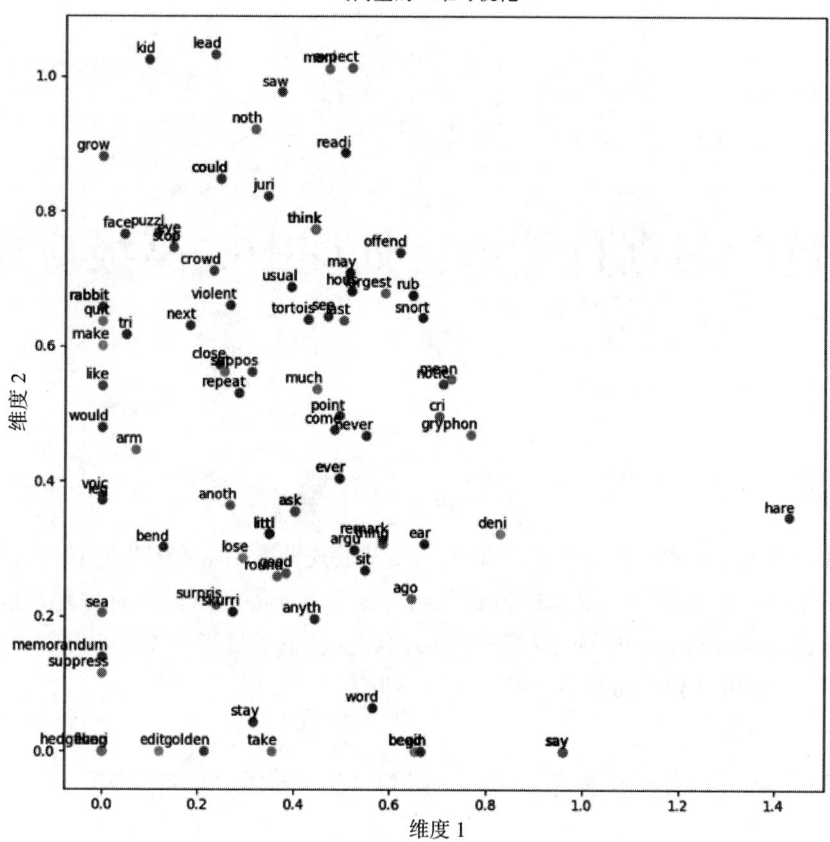
词向量的二维可视化
维度 2
维度 1
0.0
0.2
0.4
0.6
0.8
1.0
1.2
1.4
kid
lead
saw
noth
readi
grow
could
juri
think
offend
crowd
may
usual
rub
rabbit
violent
snort
next
tri
make
suppos
much
like
repeat
mean
point
cri
would
gryphon
arm
ever
anoth
ask
hare
bend
remark
ear
deni
lose
sit
ago
anyth
sea
memorandum
suppress
word
stay
edit
golden
take
say

CHAPTER 10

第10章 深度学习在自然语言处理中的高级应用

10.1 引言

深度学习技术正在被广泛地应用于解决一些开放式问题。本章我们将讨论几个不同类型的问题，这些问题都是一句简单的“是”或“否”所回答不了的具有一定难度的问题。我们希望读者能够认真学习并领会本章介绍的内容，从中了解当今深度学习领域内有哪些前沿的工作，并能够从相关的程序代码中学会一些基本构造块。

10.2 基于莎士比亚的著作使用LSTM技术自动生成文本

在本节中，我们通过使用深度**循环神经网络**（RNN）来实现基于给定的文本句子来预测下一个单词的功能。这种训练方法能够连续生成自动文本。通过在原作者的作品上进行足够的神经网络训练，生成的模型将会模仿出作者的写作风格。

10.2.1 准备工作

古登堡计划（Project Gutenberg）电子书收录了诗人威廉·莎士比亚（William Shakespeare）的所有作品，这些作品可用于训练一个自动文本生成网络。原始文本的数据资源可以从网址 http://www.gutenberg.org/ 下载：

```
>>> from __future__ import print_function
>>> import numpy as np
>>> import random
>>> import sys
```

以下代码分别创建了从索引映射到字符和从字符映射到索引的字典，这些字典用于在后续阶段将文本转换为索引信息。由于深度学习模型无法理解英文，必须要将文本转换为索引才可以实现模型训练：

```
>>> path = 'C:\\Users\\prata\\Documents\\book_codes\\ NLP_DL\\
shakespeare_final.txt'
>>&gt; text = open(path).read().lower()
>>> characters = sorted(list(set(text)))
>>> print('corpus length:', len(text))
>>> print('total chars:', len(characters))
```

```
corpus length: 581432
total chars: 61
```

```
>>> char2indices = dict((c, i) for i, c in enumerate(characters))
>>> indices2char = dict((i, c) for i, c in enumerate(characters))
```

10.2.2 如何实现

在模型训练之前，我们可以做一些预处理工作以使其效果更好，主要的步骤如下：

1. **预处理**：将整个文本文件转换成 X 和 Y 数据集，再将它们转换成索引向量化格式。
2. **深度学习模型的训练和验证**：训练和验证深度学习模型。
3. **文本生成**：用训练好的模型生成文本。

10.2.3 工作原理

下列代码展示了基于莎士比亚作品生成文本的整个建模过程。在这里我们已经设定好了字符长度。为了有效预测出下一个字符是什么，这个长度通常被设定为 40。同样地，为了避免两次连续的提取结果有重合，这个提取过程需要三个步骤来完成，这样才能创建出更好的数据集：

```
# cut the text in semi-redundant sequences of maxlen characters
>>> maxlen = 40
>>> step = 3
>>> sentences = []
>>> next_chars = []
>>> for i in range(0, len(text) - maxlen, step):
... sentences.append(text[i: i + maxlen])
... next_chars.append(text[i + maxlen])
... print('nb sequences:', len(sentences))
```

下面这张截图显示了句子总数，共计 193 798 条，这对文本生成工作来说已经足够了：

```
nb sequences: 193798
```

由于深度学习模型无法理解文本、单词和句子等，下面的代码模块将数据转换成向量格式并输入到深度学习模型。首先，将 Numpy 数组中的向量总维度值初始化为 0，然后赋值为字典的映射：

```
# Converting indices into vectorized format
>>> X = np.zeros((len(sentences), maxlen, len(characters)), dtype=np.bool)
>>> y = np.zeros((len(sentences), len(characters)), dtype=np.bool)
```

```
>>> for i, sentence in enumerate(sentences):
... for t, char in enumerate(sentence):
... X[i, t, char2indices[char]] = 1
... y[i, char2indices[next_chars[i]]] = 1
>>> from keras.models import Sequential
>>> from keras.layers import Dense, LSTM,Activation,Dropout
>>> from keras.optimizers import RMSprop
```

这里的深度学习模型采用 RNN 架构，更确切地说是带有 128 个隐藏神经元的长短期记忆网络（Long Short-Term Memory Network，LSTM），并且输出是字符级别。数组的列数即字符数。最后，在 RMSprop 优化器中调用 softmax 函数。我们也鼓励读者尝试使用其他类型的参数进行训练，最后观察训练结果有什么异同：

```
#Model Building
>>> model = Sequential()
>>> model.add(LSTM(128, input_shape=(maxlen, len(characters))))
>>> model.add(Dense(len(characters)))
>>> model.add(Activation('softmax'))
>>> model.compile(loss='categorical_crossentropy',
optimizer=RMSprop(lr=0.01))
>>> print (model.summary())
```

```
Layer (type)                 Output Shape              Param #
=================================================================
lstm_1 (LSTM)                (None, 128)               97280
_________________________________________________________________
dense_1 (Dense)              (None, 61)                7869
_________________________________________________________________
activation_1 (Activation)    (None, 61)                0
=================================================================
Total params: 105,149.0
Trainable params: 105,149.0
Non-trainable params: 0.0
_________________________________________________________________
```

如前所述，在数字索引上训练得到的深度学习模型建立了输入和输出之间的映射（给定 40 个字符，模型会预测出下一个最佳字符）。下列代码将通过模型结果找出最有可能的字符索引值，之后预测得到的索引转换回相应的字符：

```
# Function to convert prediction into index
>>> def pred_indices(preds, metric=1.0):
... preds = np.asarray(preds).astype('float64')
... preds = np.log(preds) / metric
... exp_preds = np.exp(preds)

... preds = exp_preds/np.sum(exp_preds)
... probs = np.random.multinomial(1, preds, 1)
... return np.argmax(probs)
```

通常，模型训练的的迭代次数大于 30，批尺寸（batch size）为 128。当然，读者可以自行尝试更改这些参数来查看结果的变化。代码如下：

```
# Train and Evaluate the Model
>>> for iteration in range(1, 30):
```

```
... print('-' * 40)
... print('Iteration', iteration)
... model.fit(X, y,batch_size=128,epochs=1)
... start_index = random.randint(0, len(text) - maxlen - 1)
... for diversity in [0.2, 0.7,1.2]:
... print('\n----- diversity:', diversity)
... generated = ''
... sentence = text[start_index: start_index + maxlen]
... generated += sentence
... print('----- Generating with seed: "' + sentence + '"')
... sys.stdout.write(generated)
... for i in range(400):
... x = np.zeros((1, maxlen, len(characters)))
... for t, char in enumerate(sentence):
... x[0, t, char2indices[char]] = 1.
... preds = model.predict(x, verbose=0)[0]
... next_index = pred_indices(preds, diversity)
... pred_char = indices2char[next_index]
... generated += pred_char
... sentence = sentence[1:] + pred_char
... sys.stdout.write(pred_char)
... sys.stdout.flush()
... print("\nOne combination completed \n")
```

下面这张图显示了本次训练的结果，这项结果对比了第一次迭代（Iteration 1）和最后一次迭代（Iteration 29）的差异。不难看出，只要给予模型足够的训练，文本生成效果比第一次迭代的效果要好得多：

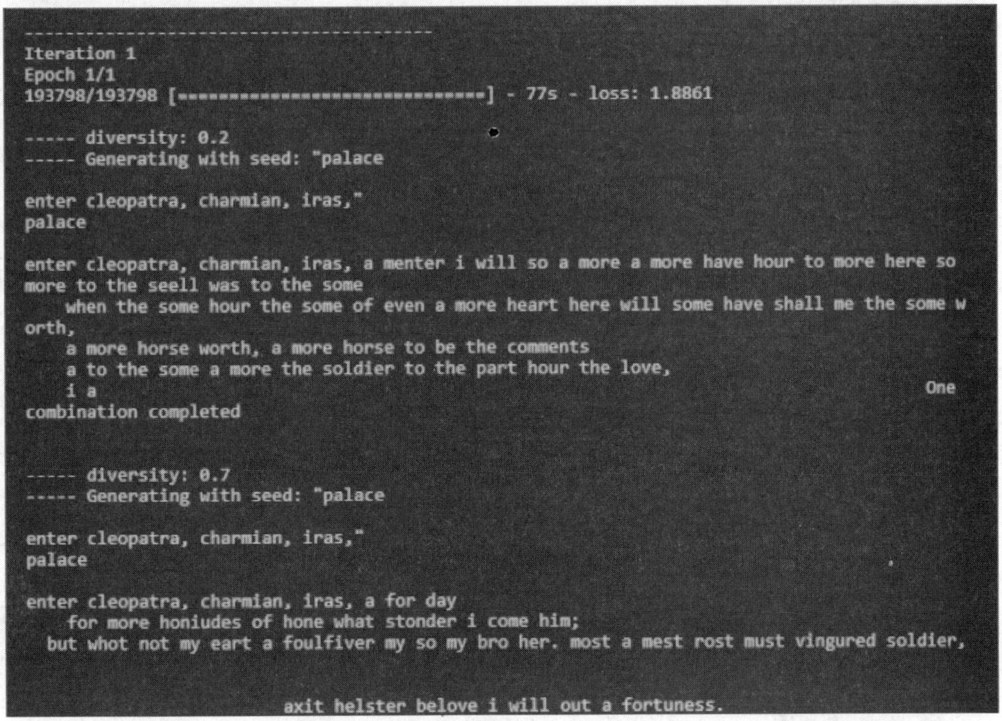

```
----------------------------------------
Iteration 1
Epoch 1/1
193798/193798 [==============================] - 77s - loss: 1.8861

----- diversity: 0.2
----- Generating with seed: "palace

enter cleopatra, charmian, iras,"
palace

enter cleopatra, charmian, iras, a menter i will so a more a more have hour to more here so
more to the seell was to the some
    when the some hour the some of even a more heart here will some have shall me the some w
orth,
    a more horse worth, a more horse to be the comments
    a to the some a more the soldier to the part hour the love,
    i a                                                                                One
combination completed


----- diversity: 0.7
----- Generating with seed: "palace

enter cleopatra, charmian, iras,"
palace

enter cleopatra, charmian, iras, a for day
    for more honiudes of hone what stonder i come him;
  but whot not my eart a foulfiver my so my bro her. most a mest rost must vingured soldier,


                          axit helster belove i will out a fortuness.
```

```
  porsest evenen dast you will not our forther but belong to that hear i mongherOne combinat
ion completed


----- diversity: 1.2
----- Generating with seed: "palace

enter cleopatra, charmian, iras,"
palace

enter cleopatra, charmian, iras, why.
i draboutur in misherss so brintow upon
    you bros! to that hath womant hons; and twreeph brountlally,
  and appercy, he your lork! beweiv a lis bitelicr,
                 ohony now fortime, pome
sake know he have whenchups, you gos, xous
   ifor mitht a
    ofverguss love?' hishwadsh, youra year trought soumshorneks, and lewas! be soleteites-
  soldients, thinksty
```

在第 29 次迭代结束后，运行结果如下：

```
--------------------------------------------------
Iteration 29
Epoch 1/1
193798/193798 [==============================] - 84s - loss: 1.2596

----- diversity: 0.2
----- Generating with seed: "to beneath your constable, it will fit
 "
to beneath your constable, it will fit
    and make the death and seem and she such counterfors,
    and i am steer place and the such so many
    the strength the way the beautions and man,
    i will not the world so straight to see him.
    what shall see thy faith the world and for my heart.
    i am so well my self with antony, and is
provided by the fortunes and madam.
  cleopatra. a prove the world with the world see the fortunesOne combination completed


----- diversity: 0.7
----- Generating with seed: "to beneath your constable, it will fit
 "
to beneath your constable, it will fit
    that we'll see him, that must thou in my excrueds,
  good of my surely slown gone to seek she grace
    of my unwore the fortunes with when what shall not out of that acceptain's war.
    but i have ages the strength such strange
    but an these fortunes, and still i do hearday shall see have you love to let us.
    your self in stedents and great the things not do friends is that be will
    One combination completed


----- diversity: 1.2
----- Generating with seed: "to beneath your constable, it will fit
 "
to beneath your constable, it will fit
    untled root silts cleopatra!
    ty haths and furethance of gentle, 'teverald once
  firthers.
    woe no, my age gait, marry of nature'd,
  and from duke ower cannot came on out yout.
atteun! dies, sir. [asidit! gender let thie, a love'
```

```
. no own mal, writh any love.
  alexas. this it hath abown of till the twongian madies.
pelo
  behaesa of well here,
  a get thee with past errike; speaking aOne combination completed
```

即便自动生成文本这项任务看起来有些不可思议，现在我们已经能够使用莎士比亚的作品来生成文本了，这也说明了，通过合理的训练和处理，我们就能模仿任意一位作家的创作风格。

10.3　基于记忆网络的情景数据问答

在本节中，我们将使用深度循环网络（deep RNN）创建一个基于情景记忆（episodic memory）的问答系统模型。该模型通过顺序阅读文章来提取给定问题的相关答案。如果要进一步获取相关内容，请参考 Ankit Kumar 等人的论文《Dynamic Memory Networks for Natural Language Processing》(https://arxiv.org/pdf/1506.07285.pdf).

10.3.1　准备工作

本节中，我们使用 Facebook 上的 BABI 数据集，该数据集可以通过 http://www.thespermwhale.com/jaseweston/babi/tasks_1-20_v1-2.tar.gz. 下载。该数据集包含大约 20 种类型的任务，我们选择了第一种任务，即构建一个单一的基于事实支持的问答系统。

将文件解压后，打开 en-10k 文件夹，将其中以 qa1_singlesupporting-fact 开头的文件作为训练和测试文件。下面的代码以特定的顺序来提取文章、问题和答案，为了生成可用于训练的数据。代码如下：

```
>>> from __future__ import division, print_function
>>> import collections
>>> import itertools
>>> import nltk
>>> import numpy as np
>>> import matplotlib.pyplot as plt
>>> import os
>>> import random
>>> def get_data(infile):
... stories, questions, answers = [], [], []
... story_text = []
... fin = open(Train_File, "rb")
... for line in fin:
... line = line.decode("utf-8").strip()
... lno, text = line.split(" ", 1)
... if "\t" in text:
... question, answer, _ = text.split("\t")
... stories.append(story_text)
... questions.append(question)
... answers.append(answer)
... story_text = []
```

```
... else:
... story_text.append(text)
... fin.close()
... return stories, questions, answers
>>> file_location = "C:/Users/prata/Documents/book_codes/NLP_DL"
>>> Train_File = os.path.join(file_location, "qa1_single-supporting-
fact_train.txt")
>>> Test_File = os.path.join(file_location, "qa1_single-supporting-
fact_test.txt")
# get the data
>>> data_train = get_data(Train_File)
>>> data_test = get_data(Test_File)
>>> print("\n\nTrain observations:",len(data_train[0]),"Test
observations:", len(data_test[0]),"\n\n")
```

提取工作完成后，我们看到训练数据集和测试数据集都生成了大约 10 000 个观察样本。如下所示：

```
Train observations: 10000 Test observations: 10000
```

10.3.2 如何实现

在基本原始集提取工作结束后，我们进行如下操作：

1. **预处理**：创建字典并将文章、问题和答案映射到词表，进一步映射成向量形式。
2. **模型开发和验证**：训练深度学习模型，然后在验证数据集的样本上进行测试。
3. **基于训练模型的预测结果**：训练模型用于预测测试数据的结果。

10.3.3 工作原理

在创建了训练数据和测试数据之后，接下来我们将依次介绍每个步骤的工作原理。

首先，建立一个词汇表的字典，其中，创建了文章、问题和答案数据中的每个词的映射。利用这些映射将单词转换成整数，然后进一步转化到向量空间中。代码如下：

```
# Building Vocab dictionary from Train and Test data
>>> dictnry = collections.Counter()
>>> for stories,questions,answers in [data_train,data_test]:
... for story in stories:
... for sent in story:
... for word in nltk.word_tokenize(sent):
... dictnry[word.lower()] +=1
... for question in questions:
... for word in nltk.word_tokenize(question):
... dictnry[word.lower()]+=1
... for answer in answers:
... for word in nltk.word_tokenize(answer):
... dictnry[word.lower()]+=1
>>> word2indx = {w:(i+1) for i,(w,_) in enumerate(dictnry.most_common() )}
>>> word2indx["PAD"] = 0
>>> indx2word = {v:k for k,v in word2indx.items()}
>>> vocab_size = len(word2indx)
>>> print("vocabulary size:",len(word2indx))
```

词汇表中的单词规模如下图所示，词汇表中只有 22 个单词，包括填充（PAD）词，它是为填充空格或零而创建的：

```
vocabulary size: 22
```

下面的代码用于确定单词的最大长度。如果我们知道了单词最大长度，就可以建立一个最大规模的向量，它可以容纳单词表中任何长度的单词：

```
# compute max sequence length for each entity
>>> story_maxlen = 0
>>> question_maxlen = 0
>>> for stories, questions, answers in [data_train,data_test]:
... for story in stories:
... story_len = 0
... for sent in story:
... swords = nltk.word_tokenize(sent)
... story_len += len(swords)
... if story_len > story_maxlen:
... story_maxlen = story_len
... for question in questions:
... question_len = len(nltk.word_tokenize(question))
... if question_len > question_maxlen:
... question_maxlen = question_len>>> print ("Story maximum
length:",story_maxlen,"Question maximum length:",question_maxlen)
```

如下图所示，文章中的单词最大长度为 14，问题中的单词最大长度为 4。对于一些文章和问题，其单词最大长度可能比上述要小，多余的维度需要用 0（或 PAD 词）来填充。这样做可以使所有的观测值具有相同的长度，使计算更加高效，否则难以实现对不同长度向量的映射或是在 GPU 中运行并行计算。

```
Story maximum length: 14 Question maximum length: 4
```

下列代码从各自的类中导入了许多函数，在后面部分我们会用到它们：

```
>>> from keras.layers import Input
>>> from keras.layers.core import Activation, Dense, Dropout, Permute
>>> from keras.layers.embeddings import Embedding
>>> from keras.layers.merge import add, concatenate, dot
>>> from keras.layers.recurrent import LSTM
>>> from keras.models import Model
>>> from keras.preprocessing.sequence import pad_sequences
>>> from keras.utils import np_utils
```

通过前面的代码我们得到了文章、问题等的单词最大长度及词表规模，在此基础上，接下来的代码段则实现了词到向量的映射：

```
# Converting data into Vectorized form
>>> def data_vectorization(data, word2indx, story_maxlen, question_maxlen):
... Xs, Xq, Y = [], [], []
... stories, questions, answers = data
... for story, question, answer in zip(stories, questions, answers):
... xs = [[word2indx[w.lower()] for w in nltk.word_tokenize(s)]
for s in story]
```

```
... xs = list(itertools.chain.from_iterable(xs))
... xq = [word2indx[w.lower()] for w in nltk.word_tokenize (question)]
... Xs.append(xs)
... Xq.append(xq)
... Y.append(word2indx[answer.lower()])
... return pad_sequences(Xs, maxlen=story_maxlen), pad_sequences(Xq,
maxlen=question_maxlen),np_utils.to_categorical(Y, num_classes=
len(word2indx))
```

下面的代码显示了 data_vectorization 的应用：

```
>>> Xstrain, Xqtrain, Ytrain = data_vectorization(data_train, word2indx,
story_maxlen, question_maxlen)
>>> Xstest, Xqtest, Ytest = data_vectorization(data_test, word2indx,
story_maxlen, question_maxlen)
>>> print("Train story",Xstrain.shape,"Train question",
Xqtrain.shape,"Train answer", Ytrain.shape)
>>> print( "Test story",Xstest.shape, "Test question",Xqtest.shape, "Test
answer",Ytest.shape)
```

下图显示了训练数据和测试数据中原文、问题和答案的向量维度：

```
Train story (10000L, 14L) Train question (10000L, 4L) Train answer (10000L, 22L)
Test story (10000L, 14L) Test question (10000L, 4L) Test answer (10000L, 22L)
```

以下代码用来初始化参数：

```
# Model Parameters
>>> EMBEDDING_SIZE = 128
>>> LATENT_SIZE = 64
>>> BATCH_SIZE = 64
>>> NUM_EPOCHS = 40
```

模型的核心代码块介绍如下：

```
# Inputs
>>> story_input = Input(shape=(story_maxlen,))
>>> question_input = Input(shape=(question_maxlen,))

# Story encoder embedding
>>> story_encoder = Embedding(input_dim=vocab_size,
output_dim=EMBEDDING_SIZE,input_length= story_maxlen) (story_input)
>>> story_encoder = Dropout(0.2)(story_encoder)

# Question encoder embedding
>>> question_encoder = Embedding(input_dim=vocab_size,output_dim=
EMBEDDING_SIZE, input_length=question_maxlen) (question_input)
>>> question_encoder = Dropout(0.3)(question_encoder)

# Match between story and question
>>> match = dot([story_encoder, question_encoder], axes=[2, 2])

# Encode story into vector space of question
>>> story_encoder_c = Embedding(input_dim=vocab_size,
output_dim=question_maxlen,input_length= story_maxlen)(story_input)
>>> story_encoder_c = Dropout(0.3)(story_encoder_c)

# Combine match and story vectors
```

```
>>> response = add([match, story_encoder_c])
>>> response = Permute((2, 1))(response)

# Combine response and question vectors to answers space
>>> answer = concatenate([response, question_encoder], axis=-1)
>>> answer = LSTM(LATENT_SIZE)(answer)
>>> answer = Dropout(0.2)(answer)
>>> answer = Dense(vocab_size)(answer)
>>> output = Activation("softmax")(answer)
>>> model = Model(inputs=[story_input, question_input], outputs=output)
>>> model.compile(optimizer="adam", loss="categorical_crossentropy",
metrics=["accuracy"])
>>> print (model.summary())
```

通过下图的模型概要，你可以了解模块的连接方式以及在模型训练中需要调参的总数：

```
Layer (type)                 Output Shape         Param #     Connected to
==========================================================================
input_1 (InputLayer)         (None, 14)           0
input_2 (InputLayer)         (None, 4)            0
embedding_1 (Embedding)      (None, 14, 128)      2816
embedding_2 (Embedding)      (None, 4, 128)       2816
dropout_1 (Dropout)          (None, 14, 128)      0
dropout_2 (Dropout)          (None, 4, 128)       0
embedding_3 (Embedding)      (None, 14, 4)        88
dot_1 (Dot)                  (None, 14, 4)        0
dropout_3 (Dropout)          (None, 14, 4)        0
add_1 (Add)                  (None, 14, 4)        0
permute_1 (Permute)          (None, 4, 14)        0
concatenate_1 (Concatenate)  (None, 4, 142)       0
lstm_1 (LSTM)                (None, 64)           52992
dropout_4 (Dropout)          (None, 64)           0
dense_1 (Dense)              (None, 22)           1430
activation_1 (Activation)    (None, 22)           0
==========================================================================
Total params: 60,142.0
Trainable params: 60,142.0
Non-trainable params: 0.0
```

下列代码实现了在训练数据上的模型拟合：

```
# Model Training
>>> history = model.fit([Xstrain, Xqtrain], [Ytrain],
batch_size=BATCH_SIZE,epochs=NUM_EPOCHS,validation_data= ([Xstest, Xqtest],
[Ytest]))
```

从第 1 轮（训练准确率为 19.35%，验证准确率为 28.98%）迭代到第 40 轮（训练准确率为 82.22%，验证准确率为 84.51%），模型的准确率有了显著提升。如下图所示：

```
Train on 10000 samples, validate on 10000 samples
Epoch 1/40
10000/10000 [==============================] - 2s - loss: 2.0210 - acc: 0.1935 - val_loss: 1.6487 - val_acc: 0.2898
Epoch 2/40
10000/10000 [==============================] - 2s - loss: 1.6443 - acc: 0.3052 - val_loss: 1.5560 - val_acc: 0.3746
Epoch 3/40
10000/10000 [==============================] - 2s - loss: 1.5162 - acc: 0.4347 - val_loss: 1.3883 - val_acc: 0.5255
Epoch 4/40
10000/10000 [==============================] - 2s - loss: 1.3762 - acc: 0.5130 - val_loss: 1.2920 - val_acc: 0.5274
Epoch 5/40
10000/10000 [==============================] - 2s - loss: 1.3155 - acc: 0.5113 - val_loss: 1.2540 - val_acc: 0.5264
Epoch 6/40
10000/10000 [==============================] - 2s - loss: 1.2806 - acc: 0.5215 - val_loss: 1.2302 - val_acc: 0.5408
Epoch 36/40
10000/10000 [==============================] - 2s - loss: 0.4343 - acc: 0.8184 - val_loss: 0.4126 - val_acc: 0.8365
Epoch 37/40
10000/10000 [==============================] - 2s - loss: 0.4357 - acc: 0.8158 - val_loss: 0.4104 - val_acc: 0.8370
Epoch 38/40
10000/10000 [==============================] - 2s - loss: 0.4297 - acc: 0.8213 - val_loss: 0.4070 - val_acc: 0.8399
Epoch 39/40
10000/10000 [==============================] - 2s - loss: 0.4311 - acc: 0.8220 - val_loss: 0.4053 - val_acc: 0.8438
Epoch 40/40
10000/10000 [==============================] - 2s - loss: 0.4260 - acc: 0.8222 - val_loss: 0.4015 - val_acc: 0.8451
```

下面的代码用于画出训练准确率及验证准确率分别随迭代次数增加的变化曲线：

```
# plot accuracy and loss plot
>>> plt.title("Accuracy")
>>> plt.plot(history.history["acc"], color="g", label="train")
>>> plt.plot(history.history["val_acc"], color="r", label="validation")
>>> plt.legend(loc="best")
>>> plt.show()
```

下图显示了准确率随迭代次数增长的变化情况，从中可以看出在第 10 次迭代之后，准确率不再大幅度上升，而是平缓上升。

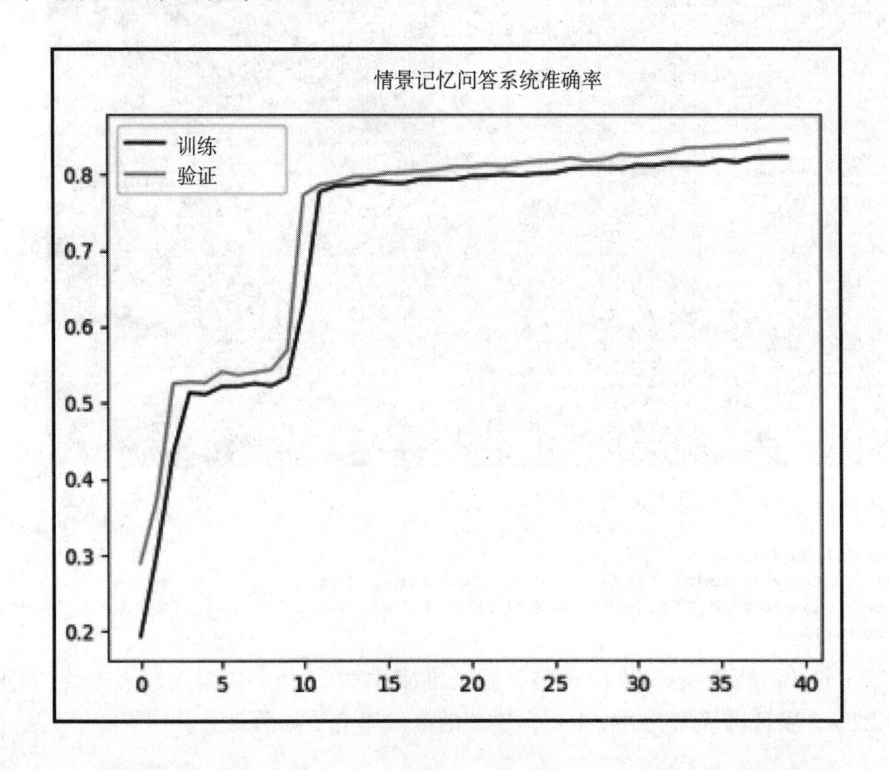

下面的代码首先获得每个相应的类的概率，之后使用 argmax 函数找到概率最大的类，该过程同时可以预测结果：

```
# get predictions of labels
>>> ytest = np.argmax(Ytest, axis=1)
>>> Ytest_ = model.predict([Xstest, Xqtest])
>>> ytest_ = np.argmax(Ytest_, axis=1)
# Select Random questions and predict answers
>>> NUM_DISPLAY = 10
>>> for i in random.sample(range(Xstest.shape[0]),NUM_DISPLAY):
... story = " ".join([indx2word[x] for x in Xstest[i].tolist() if x != 0])
... question = " ".join([indx2word[x] for x in Xqtest[i].tolist()])
... label = indx2word[ytest[i]]
... prediction = indx2word[ytest_[i]]
... print(story, question, label, prediction)
```

在对模型进行了足够的训练并在验证集上达到了较高的准确率（比如 84.51%）之后，就可以使用真实的测试数据来验证预测答案和真实答案的匹配程度了。

在 10 个随机抽取的问题中，模型仅对一个问题无法预测出正确答案（对于第 6 个问题，预测答案是 office，然而实际答案是 bedroom）。这意味着在该样本集上模型准确率为 90%。虽然我们可能无法归纳出在所有样本上该模型的准确率，但是对该模型的预测能力有了一定了解：

```
mary journeyed to the kitchen . daniel went to the bedroom . where is daniel ? bedroom bedroom
daniel went back to the hallway . sandra went to the garden . where is sandra ? garden garden
sandra journeyed to the hallway . sandra journeyed to the bathroom . where is sandra ? bathroom bathroom
john travelled to the bedroom . daniel moved to the garden . where is daniel ? garden garden
sandra journeyed to the hallway . sandra travelled to the kitchen . where is mary ? bathroom bathroom
daniel moved to the bathroom . daniel journeyed to the hallway . where is john ? bedroom office
john went to the hallway . daniel travelled to the hallway . where is daniel ? hallway hallway
john went back to the bathroom . sandra went back to the hallway . where is sandra ? hallway hallway
john went to the office . john went to the bedroom . where is john ? bedroom bedroom
john travelled to the kitchen . daniel travelled to the bathroom . where is daniel ? bathroom bathroom
```

10.4　使用循环神经网络 LSTM 进行语言建模以预测最优词

在现实世界中基于给定词来预测后续词已经有了广泛应用。举个常见例子，比如 Google 的搜索栏，当用户在搜索栏中键入自己想要搜索的关键词时，搜索栏便针对他的输入进行预测，然后给出建议。这一特点确实改善了 Google 搜索引擎的用户满意度。从技术上讲，这被称为 N 元（N-grams）模型（如果抽取了两个连续的词汇，则称之为二元（bi-grams））。目前对其建模的方法有很多，在这里基于给出的 N-1 个历史词（pre-words），我们选择深度 RNN 模型来预测后续最优词。

10.4.1　准备工作

我们使用 Alice in Wonderland（爱丽丝梦游仙境）数据完成此任务，该数据集可以从 http://www.umich.edu/~umfandsf/other/ebooks/alice30.txt 下载。在初始数据准备阶段，我们

从连续的文本文件数据中提取 N-grams，这看起来像一个故事文件：

```
>>> from __future__ import print_function
>>> import os
""" First change the following directory link to where all input files do
exist """
>>> os.chdir("C:\\Users\\prata\\Documents\\book_codes\\NLP_DL")
>>> from sklearn.model_selection import train_test_split
>>> import nltk
>>> import numpy as np
>>> import string
# File reading
>>> with open('alice_in_wonderland.txt', 'r') as content_file:
... content = content_file.read()
>>> content2 = " ".join("".join([" " if ch in string.punctuation else ch
for ch in content]).split())
>>> tokens = nltk.word_tokenize(content2)
>>> tokens = [word.lower() for word in tokens if len(word)>=2]
```

在下面的代码中，我们取 N-grams 中的 N 值为 3，这意味着每个数据片段都有 3 个连续的单词。其中，在每个数据片段中，我们都用 2 个历史词（二元）去预测下一个词。鼓励读者改变 N 值观察模型如何预测词。

注意：随着 N-grams 中的 N 值增加到 4、5 或 6 等，我们需要提供足够的增量数据来避免维数灾难。

代码如下所示：

```
# Select value of N for N grams among which N-1 are used to predict last
Nth word
>>> N = 3
>>> quads = list(nltk.ngrams(tokens,N))
>>> newl_app = []
>>> for ln in quads:
... newl = " ".join(ln)
... newl_app.append(newl)
```

10.4.2 如何实现

在提取了原始数据之后，我们需要完成下面的操作：

1. **预处理**：在预处理操作中，单词需要被转换成向量形式才能被模型处理。

2. **模型开发和验证**：建立一个将输入映射到输出的收敛 – 发散（convergent-divergent）模型，用于模型训练和验证。

3. **最优词预测**：利用训练好的模型在测试数据上预测后续最优词。

10.4.3 工作原理

使用 scikit—learn 库中的 CountVectorizer 函数将给定词（X 词和 Y 词）向量化：

```
# Vectorizing the words
>>> from sklearn.feature_extraction.text import CountVectorizer
>>> vectorizer = CountVectorizer()
>>> x_trigm = []
>>> y_trigm = []
>>> for l in newl_app:
... x_str = " ".join(l.split()[0:N-1])
... y_str = l.split()[N-1]
... x_trigm.append(x_str)
... y_trigm.append(y_str)
>>> x_trigm_check = vectorizer.fit_transform(x_trigm).todense()
>>> y_trigm_check = vectorizer.fit_transform(y_trigm).todense()
# Dictionaries from word to integer and integer to word
>>> dictnry = vectorizer.vocabulary_
>>> rev_dictnry = {v:k for k,v in dictnry.items()}
>>> X = np.array(x_trigm_check)
>>> Y = np.array(y_trigm_check)
>>> Xtrain, Xtest, Ytrain, Ytest,xtrain_tg,xtest_tg = train_test_split(X,
Y,x_trigm, test_size=0.3,random_state=42)
>>> print("X Train shape",Xtrain.shape, "Y Train shape" , Ytrain.shape)
>>> print("X Test shape",Xtest.shape, "Y Test shape" , Ytest.shape)
```

将数据转换成向量形式后，我们可以看到列值保持不变，该值表示词汇表的规模大小（图中显示词汇表的大小为 2 559）：

```
X Train shape (17947L, 2559L) Y Train shape (17947L, 2559L)
X Test shape (7692L, 2559L) Y Test shape (7692L, 2559L)
```

下面的代码是模型的核心部分，由收敛—发散体系结构组成，该体系结构可以精简和扩充神经网络的规模：

```
# Model Building
>>> from keras.layers import Input,Dense,Dropout
>>> from keras.models import Model
>>> np.random.seed(42)
>>> BATCH_SIZE = 128
>>> NUM_EPOCHS = 100
>>> input_layer = Input(shape = (Xtrain.shape[1],),name="input")
>>> first_layer = Dense(1000,activation='relu',name = "first")(input_layer)
>>> first_dropout = Dropout(0.5,name="firstdout")(first_layer)
>>> second_layer = Dense(800,activation='relu',name="second")
(first_dropout)
>>> third_layer = Dense(1000,activation='relu',name="third") (second_layer)
>>> third_dropout = Dropout(0.5,name="thirdout")(third_layer)
>>> fourth_layer = Dense(Ytrain.shape[1],activation='softmax',name =
"fourth")(third_dropout)
>>> history = Model(input_layer,fourth_layer)
>>> history.compile(optimizer = "adam",loss="categorical_crossentropy",
metrics=["accuracy"])
>>> print (history.summary())
```

下面的截图描绘了模型的完整架构，其中包含了一个收敛 – 发散结构：

Layer (type)	Output Shape	Param #
input (InputLayer)	(None, 2559L)	0

```
first (Dense)                (None, 1000)              2560000
firstdout (Dropout)          (None, 1000)              0
second (Dense)               (None, 800)               800800
third (Dense)                (None, 1000)              801000
thirdout (Dropout)           (None, 1000)              0
fourth (Dense)               (None, 2559L)             2561559
=================================================================
Total params: 6,723,359.0
Trainable params: 6,723,359.0
Non-trainable params: 0.0
```

```
# Model Training
>>> history.fit(Xtrain, Ytrain, batch_size=BATCH_SIZE,epochs=NUM_EPOCHS,
verbose=1,validation_split = 0.2)
```

在上述数据上我们对模型进行了 100 轮训练。值得注意的是，即使训练集的准确率有时有显著提升（从 5.46% 到 63.18%），然而验证集的准确率提高并不明显（从 6.63% 到 10.53%）。我们鼓励读者尝试不同的设置以进一步提高验证准确率。结果如下所示：

```
Train on 14357 samples, validate on 3590 samples
Epoch 1/100
14357/14357 [==============================] - 1s - loss: 6.3349 - acc: 0.0546 - val_loss: 6.0973 - val_acc: 0.0663
Epoch 2/100
14357/14357 [==============================] - 1s - loss: 5.9002 - acc: 0.0664 - val_loss: 6.0327 - val_acc: 0.0733
Epoch 3/100
14357/14357 [==============================] - 1s - loss: 5.6806 - acc: 0.0823 - val_loss: 5.9812 - val_acc: 0.0869
Epoch 4/100
14357/14357 [==============================] - 1s - loss: 5.4250 - acc: 0.1045 - val_loss: 5.9641 - val_acc: 0.0969
Epoch 96/100
14357/14357 [==============================] - 1s - loss: 1.1159 - acc: 0.6394 - val_loss: 9.2412 - val_acc: 0.1100
Epoch 97/100
14357/14357 [==============================] - 1s - loss: 1.1252 - acc: 0.6329 - val_loss: 9.2342 - val_acc: 0.1100
Epoch 98/100
14357/14357 [==============================] - 1s - loss: 1.1061 - acc: 0.6375 - val_loss: 9.3985 - val_acc: 0.1120
Epoch 99/100
14357/14357 [==============================] - 1s - loss: 1.1132 - acc: 0.6368 - val_loss: 9.3619 - val_acc: 0.1092
Epoch 100/100
14357/14357 [==============================] - 1s - loss: 1.1138 - acc: 0.6318 - val_loss: 9.2746 - val_acc: 0.1053
Out[11]: <keras.callbacks.History at 0xe16dc4f98>
```

```
# Model Prediction
>>> Y_pred = history.predict(Xtest)
# Sample check on Test data
>>> print ("Prior bigram words", "|Actual", "|Predicted","\n")
>>> for i in range(10):
... print (i,xtest_tg[i],"|",rev_dictnry[np.argmax(Ytest[i])],
"|",rev_dictnry[np.argmax(Y_pred[i])])
```

较低的验证准确率表示该模型可能无法很好地预测单词。造成这一问题的原因可能是特征来自于维度很高的单词级别，而不是字符级别（字符级向量维度为 26，而单词级向量维度为 2559，后者远大于前者）。下面的截图显示了两次预测的结果，显然，评价预测结果是正确（yes）还是错误（no）是非常主观的。有时预测出的单词可能不相同，但词义非常相近：

```
7148 want to | go | see
3039 neck nicely | straightened | of
2408 the rest | between | of
4068 soon finished | off | it
7093 up and | began | down
6885 her so | she | she
```

```
5985 was an | old | old
4901 have been | that | changed
4447 the roof | there | of
777 no lower | said | the
```

10.5　使用循环神经网络 LSTM 构建生成式聊天机器人

一直以来生成式聊天机器人（Generative chatbot）的构建是一个相当困难的任务。即使在今天，大多数可使用的聊天机器人在本质上都依赖于检索机制。对于给定的问题，它们基于语义相似度和意图等线索检索出最佳回复。要想进一步获取有关内容，请参阅 Kyunghyun Cho 等人的论文《Learning Phrase Representations using RNN Encoder-Decoder for Statistical Machine Translation》(https://arxiv.org/pdf/1406.1078.pdf。

10.5.1　准备工作

使用 A.L.I.C.E 人工智能基金会的人工智能标注语言数据集 bot.aiml（Artificial Intelligence Markup Language，AIML）来训练模型，该数据集有自定义语法，比如 XML 文件。在该文件中，问题和答案是存在映射的，每个问题都有特定的答案与之对应。完整的 bot.aiml 数据集可以在 https://code.google.com/archive/p/aiml-en-us-foundation-alice/downloads 的 aiml-en-us-foundation-alice.v1-9 文件中获得。解压文件夹并找到 bot.aiml 文件，然后使用 Notepad 将其打开。接着将其保存为 bot.txt 以便在 Python 中读入：

```
>>> import os
""" First change the following directory link to where all input files do
exist """
>>> os.chdir("C:\\Users\\prata\\Documents\\book_codes\\NLP_DL")
>>> import numpy as np
>>> import pandas as pd
# File reading
>>> with open('bot.txt', 'r') as content_file:
... botdata = content_file.read()
>>> Questions = []
>>> Answers = []
```

与 XML 文件相似，AIML 文件也有其独特语法。其中，pattern 表示问题，template 表示答案。因此，我们分别进行提取：

```
>>> for line in botdata.split("</pattern>"):
... if "<pattern>" in line:
... Quesn = line[line.find("<pattern>")+len("<pattern>"):]
... Questions.append(Quesn.lower())
>>> for line in botdata.split("</template>"):
... if "<template>" in line:
... Ans = line[line.find("<template>")+len("<template>"):]
... Ans = Ans.lower()
... Answers.append(Ans.lower())
```

```
>>> QnAdata = pd.DataFrame(np.column_stack([Questions,Answers]),columns =
["Questions","Answers"])
>>> QnAdata["QnAcomb"] = QnAdata["Questions"]+" "+QnAdata["Answers"]
>>> print(QnAdata.head())
```

由于我们需要将所有的单词/字符转换成数值表示，因此，我们把问题和答案合并起来以提取出在模型中使用的词汇表。这样做的原因与前一小节相同——深度学习模型无法处理英文，所有的内容必须转换成数值形式才能被模型处理。

```
            Questions                                              Answers  \
0               yahoo  a lot of people hear about <bot name="name"/> ...
1        you are lazy                      actually i work 24 hours a day.
2         you are mad                    no i am quite logical and rational.
3    you are thinking  i am a thinking machine.<think><set name="it">...
4  you are dividing *            actually i am not too good at division.

                                             QnAcomb
0  yahoo a lot of people hear about <bot name="na...
1       you are lazy actually i work 24 hours a day.
2     you are mad no i am quite logical and rational.
3  you are thinking i am a thinking machine.<thin...
4  you are dividing * actually i am not too good ...
```

10.5.2 如何实现

在提取问答对之后，完成以下步骤来处理数据并得到结果：

1. **预处理**：将问答对转换成向量形式，用于模型训练。
2. **模型建立和验证**：构建深度学习模型并验证数据。
3. **模型训练用于预测答案**：训练模型，利用模型预测给定问题的答案。

10.5.3 工作原理

问题和答案数据用来创建词到索引的映射词汇表，该表将用于词到向量的转换：

```
# Creating Vocabulary
>>> import nltk
>>> import collections
>>> counter = collections.Counter()
>>> for i in range(len(QnAdata)):
... for word in nltk.word_tokenize(QnAdata.iloc[i][2]):
... counter[word]+=1
>>> word2idx = {w:(i+1) for i,(w,_) in enumerate(counter.most_common())}
>>> idx2word = {v:k for k,v in word2idx.items()}
>>> idx2word[0] = "PAD"
>>> vocab_size = len(word2idx)+1
>>> print (vocab_size)
```

```
Vocabulary size: 3451
```

下面的编码函数用于将文本转换成索引，解码函数用于将索引转换回文本。众所周知，深度学习模型只能处理数值型数据，不能处理文本或字符型数据：

```
>>> def encode(sentence, maxlen,vocab_size):
... indices = np.zeros((maxlen, vocab_size))
... for i, w in enumerate(nltk.word_tokenize(sentence)):
... if i == maxlen: break
... indices[i, word2idx[w]] = 1
... return indices
>>> def decode(indices, calc_argmax=True):
... if calc_argmax:
... indices = np.argmax(indices, axis=-1)
... return ' '.join(idx2word[x] for x in indices)
```

给定了问题和答案的最大长度后，下列代码将问题和答案进行了向量化。问题和答案的最大长度可能会不同。在一些数据中，问题的长度要大于答案的长度；而在少数情况下，前者要小于后者。理论上问题更长对捕捉正确答案很有帮助。然而遗憾的是，在本例中问题长度比答案长度要短得多，这对开发生成式模型来说并不是一个好样本：

```
>>> question_maxlen = 10
>>> answer_maxlen = 20
>>> def create_questions(question_maxlen,vocab_size):

... question_idx = np.zeros(shape=(len(Questions),question_maxlen,
vocab_size))
... for q in range(len(Questions)):
... question = encode(Questions[q],question_maxlen,vocab_size)
... question_idx[i] = question
... return question_idx
>>> quesns_train = create_questions(question_maxlen=question_maxlen,
vocab_size=vocab_size)
>>> def create_answers(answer_maxlen,vocab_size):
... answer_idx = np.zeros(shape=(len(Answers),answer_maxlen, vocab_size))
... for q in range(len(Answers)):
... answer = encode(Answers[q],answer_maxlen,vocab_size)
... answer_idx[i] = answer
... return answer_idx
>>> answs_train = create_answers(answer_maxlen=answer_maxlen,vocab_size=
vocab_size)
>>> from keras.layers import Input,Dense,Dropout,Activation
>>> from keras.models import Model
>>> from keras.layers.recurrent import LSTM
>>> from keras.layers.wrappers import Bidirectional
>>> from keras.layers import RepeatVector, TimeDistributed,
ActivityRegularization
```

聊天机器人的关键代码如下所示。这里我们使用了循环神经网络、重复向量和时间分布式网络。重复向量用于输入维度和输出值的匹配，而时间分布式网络则用于将列向量改变为输出维度的词表大小：

```
>>> n_hidden = 128
>>> question_layer = Input(shape=(question_maxlen,vocab_size))
>>> encoder_rnn = LSTM(n_hidden,dropout=0.2,recurrent_dropout=0.2)
(question_layer)
>>> repeat_encode = RepeatVector(answer_maxlen)(encoder_rnn)
```

```
>>> dense_layer = TimeDistributed(Dense(vocab_size))(repeat_encode)
>>> regularized_layer = ActivityRegularization(l2=1)(dense_layer)
>>> softmax_layer = Activation('softmax')(regularized_layer)
>>> model = Model([question_layer],[softmax_layer])
>>> model.compile(loss='categorical_crossentropy', optimizer='adam',
metrics=['accuracy'])
>>> print (model.summary())
```

下面的模型概要描述了模型规模大小的变化。输入层与问题向量的维度相匹配，而输出层与答案向量的维度相匹配：

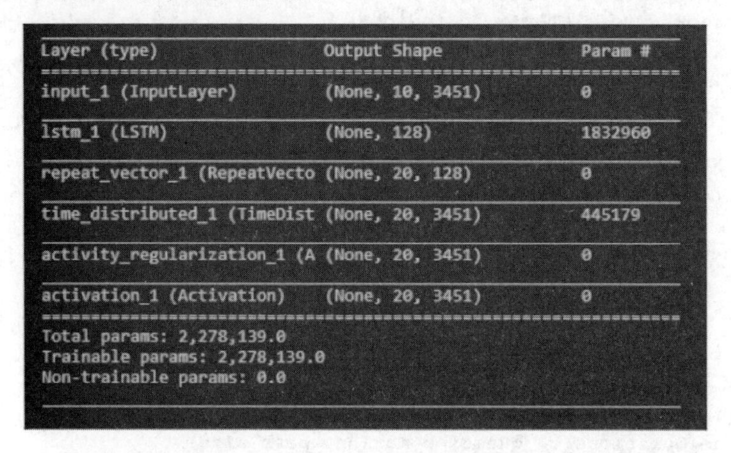

Layer (type)	Output Shape	Param #
input_1 (InputLayer)	(None, 10, 3451)	0
lstm_1 (LSTM)	(None, 128)	1832960
repeat_vector_1 (RepeatVecto	(None, 20, 128)	0
time_distributed_1 (TimeDist	(None, 20, 3451)	445179
activity_regularization_1 (A	(None, 20, 3451)	0
activation_1 (Activation)	(None, 20, 3451)	0

Total params: 2,278,139.0
Trainable params: 2,278,139.0
Non-trainable params: 0.0

```
# Model Training
>>> quesns_train_2 = quesns_train.astype('float32')
>>> answs_train_2 = answs_train.astype('float32')
>>> model.fit(quesns_train_2, answs_train_2,batch_size=32,epochs=30,
validation_split=0.05)
```

结果如下图所示，虽然准确率很高，但结果却不容乐观。聊天机器人模型可能生成毫无意义的文字，答非所问。为什么呢？产生该问题的原因是数据中单词数量过少。

```
Train on 2803 samples, validate on 148 samples
Epoch 1/30
2803/2803 [==============================] - 3s - loss: 0.0000e+00 - acc: 1.0000 - val_loss: 0.0571 - val_acc: 0.9932
Epoch 2/30
2803/2803 [==============================] - 3s - loss: 0.0000e+00 - acc: 1.0000 - val_loss: 0.0571 - val_acc: 0.9932
Epoch 3/30
2803/2803 [==============================] - 3s - loss: 0.0000e+00 - acc: 1.0000 - val_loss: 0.0571 - val_acc: 0.9932
Epoch 4/30
2803/2803 [==============================] - 3s - loss: 0.0000e+00 - acc: 1.0000 - val_loss: 0.0571 - val_acc: 0.9932
Epoch 26/30
2803/2803 [==============================] - 3s - loss: 0.0000e+00 - acc: 1.0000 - val_loss: 0.0571 - val_acc: 0.9932
Epoch 27/30
2803/2803 [==============================] - 3s - loss: 0.0000e+00 - acc: 1.0000 - val_loss: 0.0571 - val_acc: 0.9932
Epoch 28/30
2803/2803 [==============================] - 3s - loss: 0.0000e+00 - acc: 1.0000 - val_loss: 0.0571 - val_acc: 0.9932
Epoch 29/30
2803/2803 [==============================] - 3s - loss: 0.0000e+00 - acc: 1.0000 - val_loss: 0.0571 - val_acc: 0.9932
Epoch 30/30
2803/2803 [==============================] - 3s - loss: 0.0000e+00 - acc: 1.0000 - val_loss: 0.0571 - val_acc: 0.9932
Out[13]: <keras.callbacks.History at 0x17efbe358>
```

```
# Model prediction
>>> ans_pred = model.predict(quesns_train_2[0:3])
>>> print (decode(ans_pred[0]))
>>> print (decode(ans_pred[1]))
```

下面的截图显示了测试数据上的输出样例。不难看出，输出结果毫无意义，这是生成式模型导致的问题。

```
PAD PAD PAD PAD PAD PAD PAD PAD PAD PAD PAD PAD PAD PAD PAD PAD PAD PAD PAD PAD
PAD PAD PAD PAD PAD PAD PAD PAD PAD PAD PAD PAD PAD PAD PAD PAD PAD PAD PAD PAD
```

我们的模型在这种情况下效果不佳，但是生成式聊天机器人模型中仍然有一些可以改进的地方。读者可以做如下尝试：

- 采用长度更长的问题和答案数据集，以便更好地获取信息。
- 创建一个规模相对较大的深度学习模型架构，并执行更多轮的迭代训练。
- 采用通用型的问答对，比如检索知识等，而不是虚构的数据。这种数据难以生成模型。

推荐阅读

Deep Learning
Core Technology, Tools and Case Study
深度学习
核心技术、工具与案例解析

Deep Learning in Action
深度学习
主流框架和编程实战

自然语言处理与深度学习
通过C语言模拟

Java Deep Learning Essentials
深度学习
Java语言实现

O'REILLY
TensorFlow
学习指南
深度学习系统构建详解
Tom Hope Yehezkel S. Resheff Itay Lieder 著

Deep Learning by PaddlePaddle
PaddlePaddle
深度学习实战

Neural Network Methods
for Natural Language Processing
基于深度学习的
自然语言处理

Natural Language Processing
Python自然语言处理实战
核心技术与算法

Deep Convolutional Neural Network
Principle and Practice
深度卷积网络
原理与实践